# The IMA Volumes
# in Mathematics
# and Its Applications

## Volume 14

*Series Editors*
Hans Weinberger  Willard Miller, Jr.

# Institute for Mathematics and Its Applications
# IMA

The **Institute for Mathematics and Its Applications** was established by a grant from the National Science Foundation to the University of Minnesota in 1982. The IMA seeks to encourage the development and study of fresh mathematical concepts and questions of concern to the other sciences by bringing together mathematicians and scientists from diverse fields in an atmosphere that will stimulate discussion and collaboration.

The IMA Volumes are intended to involve the broader scientific community in this process.

Hans Weinberger, Director
Willard Miller, Jr., Associate Director

## IMA Programs

1982–1983 Statistical and Continuum Approaches to Phase Transition

1983–1984 Mathematical Models for the Economics of Decentralized Resource Allocation

1984–1985 Continuum Physics and Partial Differential Equations

1985–1986 Stochastic Differential Equations and Their Applications

1986–1987 Scientific Computation

1987–1988 Applied Combinatorics

1988–1989 Nonlinear Waves

1989–1990 Dynamical Systems and Their Applications

## Springer Lecture Notes from the IMA

*The Mathematics and Physics of Disordered Media*
Editors: Barry Hughes and Barry Ninham
(Lecture Notes in Mathematics, Volume 1035, 1983)

*Orienting Polymers*
Editor: J. L. Ericksen
(Lecture Notes in Mathematics, Volume 1063, 1984)

*New Perspectives in Thermodynamics*
Editor: James Serrin
(Springer-Verlag, 1986)

*Models of Econoic Dynamics*
Editor: Hugo Sonnenschein
(Lecture Notes In Economics, Volume 264, 1986)

J.R. Rice
Editor

# Mathematical Aspects of
# Scientific Software

With 46 Illustrations

Springer-Verlag
New York  Berlin  Heidelberg
London  Paris  Tokyo

J.R. Rice
Department of Computer Science
Purdue University
West Lafayette, IN 47907, USA

Mathematics Subject Classification (1980): 68Q20

Library of Congress Cataloging-in-Publication Data
Mathematical aspects of scientific software.
   (The IMA volumes in mathematics and its
applications; v. 14)
   Bibliography: p.
   1. Computer software—Development.   2. Science—
Data processing.   I. Rice, John Richard.      II. Series.
QA76.76.D47M366   1988      502'.85'53         88-3237

Camera-ready text prepared by the editor.

9 8 7 6 5 4 3 2 1

ISBN-13: 978-1-4684-7076-5      e-ISBN-13: 978-1-4684-7074-1
DOI: 10.1007/978-1-4684-7074-1

# The IMA Volumes in Mathematics and Its Applications

## Current Volumes:

**Forthcoming Volumes:**

1986–1987: *Scientific Computation*

The Modeling of Fractures, Heterogeneities, and Viscous Fingering in Flow in Porous Media

Computational Fluid Dynamics and Reacting Gas Flows

Numerical Algorithms for Modern Parallel Computer Architectures

Atomic and Molecular Structure and Dynamics

# CONTENTS

# FOREWORD

This IMA Volume in Mathematics and its Applications

## MATHEMATICAL ASPECTS OF SCIENTIFIC SOFTWARE

is in part the proceedings of a workshop which was an integral part of the 1986-87 IMA program on SCIENTIFIC COMPUTATION. We are grateful to the Scientific Committee: Bjorn Engquist (Chairman), Roland Glowinski, Mitchell Luskin and Andrew Majda for planning and implementing an exciting and stimulating year-long program. We especially thank the Workshop Organizer, John R. Rice for organizing a workshop which brought together many of the major figures in a variety of research fields connected with scientific software for a fruitful exchange of ideas.

Willard Miller, Jr.

Hans Weinberger

# PREFACE

Scientific software is the fuel that drives today's computers to solve a vast range of problems. Huge efforts are being put into developing new software, new systems and new algorithms for scientific problem solving. The ramifications of this effort echo throughout science and, in particular, into mathematics. This book explores how scientific software impacts the structure of mathematics, how it creates new subfields and how new classes of mathematical problems arise.

The focus is on five topics where the impact is currently being felt and where important new challenges for mathematics exist. These topics are the new subfields of parallel and geometric computations, the emergence of symbolic computation systems into "general" use, the potential emergence of new, high-level mathematical systems, and the crucial question of how to measure the performance of mathematical problem solving tools.

This workshop brought together research workers from the borders between mathematics and computer science, people who were able to see and discuss the interactions between mathematics and its applications. The editor greatly appreciates the efforts made by the authors and other participants in the workshop. Special thanks are due to Carl de Boor, Clarence Lehman, Bradley Lucier and Richard McGehee who gave stimulating and insightful panel presentations on the nature and needs for high level mathematical systems.

# MATHEMATICAL ASPECTS OF SCIENTIFIC SOFTWARE

JOHN R. RICE*

Department of Computer Science

Purdue University

West Lafayette, Indiana 47907

## Abstract

The goal is to survey the impact of scientific software on mathematics. Three types of impacts are identified and several topics from each are discussed in some depth. First is the impact on the structure of mathematics through its role as the scientific tool for problem solving. Scientific software leads to new assessments of what algorithms are, how well they work, and what a solution really is. Second is the initiation of new mathematical endeavors. Numerical computation is already very widely known, we discuss the important future roles of symbolic and geometric computation. Finally, there are particular mathematical problems that arise from scientific software. Examples discussed include round-off errors and the validation of computations, mapping problems and algorithms into machines, and adaptive methods. There is considerable discussion of the shortcommings of mathematics in providing an adequate model for the scientific analysis of scientific software.

## 1. The Impact of Scientific Software on Mathematics

The goal of this paper is to survey the impact of scientific software on mathematics. Three areas are identified:

1) *The effect on the structure of mathematics*, on what mathematicians do and how they view their activities,

2) *New mathematical endeavors that arise*, new specialities or subspecialities of mathematics that may be created,

---

*This work supported in part by the Air Force Office of Scientific Research grant 84-0385.

3)   *Mathematical problems that arise,* difficult mathematical problems or groups
     of problems arise from efforts to understand certain methods or phenomena of
     scientific software.

About 15 topics are presented which illustrate these impacts. No attempt has been made
to be encyclopedic, the choices are those that appeal to the author. This introductory section is a summary of the survey, about a dozen of the topics are discussed in more depth
in the later sections of this paper and only mentioned here. A few topics not discussed
later are included here with a few remarks.

In considering the structure of mathematics, it is important to realize that not only
does mathematics grow but that it also changes nature. Large subfields die out and not
just because all the problems are solved or questions answered. There subfields become
irrelevant to new directions that mathematics take. An example of this exists in scientific
computation, namely making tables of mathematical functions. This endeavor started
almost in the antiquity of mathematics and grew until there were a large number of practitioners. Volumes and volumes of tables were prepared and the methodology of creating, checking and making them easy to use became quite sophisticated. This endeavor is
now in a steep decline because computers and scientific software have made it easier and
more reliable to compute values from "first principles" than to look them up in tables.

Scientific software will lead mathematics to focus again more heavily on *problem
solving and algorithms.* We need to analyze the intrinsic nature of problems, how hard
they are to solve and the strengths of classes of algorithms. We need to examine again
what it means to solve a problem, a step that will show many previous "solutions" to be
of little value.

We need to examine again what it means to prove a result, and what techniques are
reliable. Some groups in computer science have a substantially different view of proof
then modern mathematical practice. They view proofs much more formally, somewhat
reminiscent of Russell and Whitehead's Principia Mathematica. If computer programs
are going to prove theorems, how does one prove the programs themselves are correct?
For a delightful and insightful analysis of the pitfalls here, see [Davis, 1972]. I believe
that we have learned several important things about proofs and scientific software. First,
scientific software is not amenable to proofs as a whole because it contains many heuristics (i.e., algorithm fragments for which there is no underlying formal model, we just
hope they work). Second, it is very difficult, often impossible, to say what scientific

software is supposed to do. Finally, the computing requirements for this approach are truly enormous.

We discuss the impact on mathematics of the creation and widespread use of *mathematical systems*. This development is now overdue and will have an enormous impact on education and practice of mathematics. It is plausible that one can automate large parts (the algorithmic parts) of mathematics from the middle elementary years to the middle undergraduate years. The results will be much more reliable abilities, less cost, more power for problem solving and more time to learn the mysteries of problem solving rather than the rote.

The largest and most visible new mathematical endeavor resulting from scientific software is that of *numerical computation*. It has a lot of structure and activity. The principle components are *numerical analysis*, a large subfield of mathematics and computer science, *mathematical software*, a smaller subfield of computer science, a part of *applied mathematics*, and *computational analysis*, a huge subfield of science and engineering. This endeavor has not yet had a large impact on most of mathematics. Perhaps this is because mathematics has turned away from problem solving and has been content to let other disciplines appropriate this endeavor. As the attention of mathematics is turned toward problem solving, interest in this area will rise substantially.

Sections 6 to 8 discuss the newer endeavors of *symbolic computation* and *geometric computation*. These are currently much smaller than numerical computation but may have a more immediate impact on mathematics because they are closer to currently active areas. Geometric computation is still quite immature and offers a host of challenges for mathematics.

Perhaps the best known problem area arising from scientific software is that of *round-off error analysis*. In 1948 John von Newmann and Herbert Goldstine pursued the idea of following the effect of individual errors and bounding the results. This is very tedious and overall this approach is a failure. A second and more subtle idea is to estimate the change in the problem data so the computed result is exact. This is often very practical and it gives very useful information, once one accepts that a realistic estimate of the error due to round-off is not going to be obtained. The third and most widely used idea is to do all computations with much higher precision than is thought to be required. One can even use exact arithmetic. This approach is quite, but not extremely, reliable. It is also expensive. The final approach has been to introduce *condition numbers*, these are essentially norms of the Frechet derivative of the solution with respect to the data. Note that these do not take into account actual round-off errors but rather estimate the uncertainty of the solution in terms of the uncertainty of the problem data.

Round-off error analysis is somewhat unpopular because it is so frustrating and many people hope they can succeed in ignoring it. That is the real attraction of buying computers with long (e.g., 64 bit) word lengths, one gets high precision and, hopefully, freedom from worrying about round-off. Unfortunately there is a general lack of understanding of the nature of uncertainty effects in problem solving. There is confusion about the difference between the condition of a problem and that of an algorithm to solve it. If a problem is badly conditioned (the condition number is large) then nothing can be done about it while a badly conditioned algorithm might be replaced by a better one.

Condition numbers are sometimes misleading in that the estimates derived are grossly pessimistic. In my own work of solving elliptic PDEs, I see condition numbers like $10^5$, $10^{10}$ or $10^{15}$ and yet observe almost no round-off effects. This leads to more confusion which is further compounded by the fact that scaling problems (simply changing the units of measurement) can have dramatic effects on round-off. This phenomena is poorly understood and difficult to analyze.

The problem of round-off error is just one aspect of a more general question: *How do you know the computed results are correct?* The mathematical results here are much less than satisfactory. Most problems addressed by scientific software are unsolvable within the framework of current mathematics. For example, given a program to compute integrals numerically, it is easy to construct a function (with as many derivatives as one wants) where the result is as inaccurate as one wants. Most theorems that apply to prove correctness of computed results have unverifiable hypotheses. And many theorems have hypotheses that are obviously violated in common applications. Most scientific software contains several heuristic code fragments. Indeed, it is usually not possible to give a precise mathematical statement of what the software is supposed to do.

The search for techniques to give better confidence in computed results is still on. A posteriori techniques still are not fully explored (it is easy to tell if $x_o$ solves $f(x) = 0$). Computing multiple solutions efficiently is another technique that holds promise and which uses the old idea: Solve the problem 3 times (or $k$ times) and with 3 methods and compare the results. The application of several techniques is usually required to achieve really high confidence in correctness. A rule of thumb is that it costs as much to verify the correctness of a computed result as to compute it in the first place.

Four other topics are discussed in Sections 9 to 12: 1) mapping problems and algorithms into the new parallel machines, 2) the analysis of adaptive algorithms, 3) how well mathematics models real problems, can one find theorems that are useful in assessing real computations, 4) the role mathematics plays in the experimental performance evaluation of scientific software. The final topic is particularly frustrating. Much like the weather, everyone talks about it but few do anything about it. One frequently hears statements "method $x$ is the best way to solve problem $y$" which are in fact, little more than conjectures. Scientific and systematic performance evaluation is a lot of work, much of it is tedious and the work is not highly regarded by one's peers. No wonder that people prefer to do other things. We have the puzzling situation where research managers and funding agencies are always looking for "better" methods and yet they are uninterested in supporting work to measure which methods are actually good or bad.

## 2. Problem Solving and Algorithms

Historically, mathematics has arisen from the need to solve problems. It was recognized about a thousand years ago that one can codify the steps needed to solve some problems. These steps can be written down and someone can be told *"If you have this kind of problem, then follow these steps and you will have the solution."* This idea matured into two of the most fundmental concepts inn mathematics: *models* and *algorithms*. Models arise from the need to make precise the phrase *"this kind of problem"* and algorithms make precise the phrase *"follow these steps"*. Recall that, intuitively speaking, a model is an abstract system using axioms, assumptions and definitions which represents (well, one hopes) a real would system. An algorithm is a set of precise instructions to operate an abstract machine.

Mathematics has evolved through several well identified levels of problem solving. The lowest several of these are presented below along with typical problems and solutions.

## Arithmetic

| Problem | Solution |
| --- | --- |
| What is $2 + 2$? | 4 |
| What is $7 \times 8$? | 56 |
| What is $1327/83$? | $15.9879518...$ |
| What is $3/2 \times (1/8 - 3/5 + 1/12) / (37/8)$? | $47/310$ |
| What is $\sqrt{86}$? | $9.273618...$ |

*Notes.* There are many algorithms including memorization (table look-up in computer science terms) taught in school. It is significant that some studies suggest that about 80% of the entering college freshman cannot do long division, i.e., do not know an algorithm for computing 1327/83 as a decimal number. I would guess that a greater percentage of professional mathematicians and scientists cannot take square roots.

## Algebra

| Problem | Solution |
| --- | --- |
| What is $3x + 2y - x - 3y$? | $2x - y$ |
| What is $(3x + 2y) \times (x + 3y)$? | $3x^2 + 11xy + 6y^2$ |
| Solve $3x^2 - x - 7 = 0$ | $x = 1.703...$ |
| Solve $x^3 - 7x^2 + 3x - 110 = 0$ | $x = 8.251...$ |

*Note.* Very few mathematicians know the algorithms for all of these problems.

## Calculus

| Problem | Solution |
| --- | --- |
| What is the derivative of $e^x$? | $e^x$ |
| What is the integral of $\cot x$? | $\log|\sin x| + c$ |
| What is the integral of $(\cos x)/x$? | $\log|x| + c - \sum_{i=1}^{\infty} \dfrac{(-1)^i x^{2i}}{2i(2i)!}$ |
| What is the area under the curve $y = 1/(\sqrt{x}(1 + x))$ for $x$ in $[0,\infty]$? | $\pi$ |
| What is the series expansion of $erf(x)$ | $\dfrac{2\pi}{\sqrt{x}} \sum_{i=0}^{\infty} \dfrac{(-1)^i x^{2i}}{(2i + 1)i!}$ |

*Notes.* The algorithms learned in calculus are rarely explicitly stated or taught. Problems are often called "solved" when one symbolic expression symbolic expression (say, a function or integral) is shown equal to another symbolic expression (say, an infinite series or product), neither of which can be computed exactly (even assuming one can do exact arithmetic with real numbers).

Beyond these three levels there are linear algebra, ordinary differential equations, combinatorics and others.

Even though the essense of mathematics is model construction and problem solving, a large percentage of the teaching effort is devoted to learning algorithms (often with neither the teacher or student being aware of this). I believe that it is much more important to *learn how to use knowledge* than to *learn knowledge*. In mathematical terms, it is more important to learn how to use algorithms than to memorize them.

The current situation is as follows:

(a)   Enormous effort is invested in teaching people algorithms.

(b)   People forget most of the algorithms they learn.

(c)   Many algorithms of arithmetic, algebra, calculus, linear algebra, etc., can be implemented as scientific software and run on cheap machines.

(d)   Many educators who expound the virtues of learning algorithms routinely use concepts, processes and models for which they do not know any relevent algorithms.

This situation is unstable and portends great changes in the educational system. This change will be slow but profound. Almost twenty years ago a compute program could make a grade of B on some calculus exams at MIT. Surely we cannot continue this when the cost of machines and software to do algorithms is becoming negligible.

It is fascinating to contemplate what the first two years of college mathematics would be if it were based on a mathematical system which includes the standard (and not so standard) algorithms of arithmetic, algebra, calculus, linear algebra, numerical analysis, geometry and combinatorics. The National Academy of Sciences is sponsoring a new study *Calculus for a New Century*, perhaps it will make some steps of change.

## 3.   How Hard are Problems to Solve?

Since problem solving is one focus of mathematics, a central question is to determine just how hard various problems are to solve. Being hard to solve is measured by how much computation an algorithm must do, not by how hard it is to discover the algorithm. The problem of multiplying two integers, compute $a \times b$, illustrates the idea.

The "machine" to be used can multiply single digit integers and do addition of integers. The size of the problem is measured by the length $N$ of $a$ and $b$. The schoolboy algorithm requires $N^2$ single digit multiplications (every digit of $a$ is multiplied by every digit of $b$) and $N$ additions of long integers. This problem is thus possibly $O(N^2)$ hard, there might be no faster way to compute the product. Some thought leads to a method that only requires $O(N \log N)$ operations, it is much harder to show that it cannot be done with $O(N)$ operations (of standard abstract computing machines). So we know the intrinsic difficulty of computing $a \times b$ is not quite linear, $O(N)$, in the size of the problem and that the schoolboy algorithm is $O(N^2)$ but easy to remember.

There are some problems, often of a combinatorial nature, that are exponentially hard, i.e., the intrinsic work to solve these problems grows exponentially with the size of the problem. An example of such a problem is the *traveling salesman problem* where one is to compute the shortest route that visits each one of a set of $N$ cities.

We give three practical questions in this area along with a few remarks.

1) *Are linear programming problems exponentially hard to solve?*

The simplex algorithm routinely solves these problems with $N$ variables and constraints in roughly $O(N)$ steps. Yet an example has been discovered (after many years of search) where the simplex algorithm takes $O(2^N)$ steps. More recently, the discovery of algorithms for these problems whose worst case is only $O(N^7)$ or $O(N^3)$ or ?? has created somewhat of a sensation in the press. It is still not yet clear how hard these problems really are.

2) *Are there algorithms to solve partial differential equations (PDEs) that are as efficient as evaluating a closed form solution?*

The theory of PDEs is one of the most sophisticated in mathematics and PDEs have long been regarded as among the most difficult problems to solve in applied mathematics. Dramatic progress [see, Rice, 1983, Chapter 10] has been made in developing better algorithms to solve PDEs and there is growing evidence that the answer to this question may be yes.

3) *Are systems of non-linear equations exponentially hard to solve?*

One sees examples of systems of thousands of equations being solved and yet one can construct rather simple, apparently well behaved systems with 10 variables that defeat all known algorithms. It may be that lots of large real-world problems are feasible to solve while the problem class as a whole is intractable. If so, it is of great interest to discover why so many real-world problems are "easy" to solve.

There are two similar mathematical frameworks for studying this problem. The best known one is *complexity theory*. See, for example, [Aho, Hopcroft and Ullman, 1974]. A simple model of this theory is illustrated by the following schematic for problem solving:

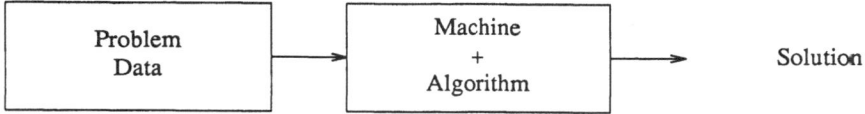

The central question here is: *for a given machine and problem class, find an algorithm which produces the solution with the fewest steps.* Simple examples of this theory are listed below (we are the notation that $N$ is an integer, $x$ is a real, $f$ is a function, $A$ is a matrix, and $x, b$ are vectors):

| Problem | Data | Machine | Solution Type |
|---|---|---|---|
| Evaluate $f(N)$ | $N$ | Integer arithmetic (Turing machine) | integer |
| Multiply $a \times b$ | $a, b$ N-digits | Addition 1 digit multiply | integer |
| Solve $A x = b$ | matrix A vector $b$ order $N$ | Real arithmetic (Fortran) | vector of reals |
| Evaluate $f(x)$ to $N-digits$ | real $x$ $N-digits$ | Real arithmetic | real |

These four examples illustrate three facts about this theory: 1) The "power" of the machine must be carefully controlled (if the machine can multiply reals, then $a \times b$ is trivial). 2) The problem class must be carefully specified (if $a$ and $b$ are reals, then the problem of multiplying $a \times b$ is unsolvable with a machine that can only do single digit multiplication and addition). 3) It is nice to have one or two simple parameters that characterize the "size" of the problem (e.g., number of digits in input, order of matrix and vectors in input, number of digits in output).

The second framework is that of *approximation theory* (see, for example, [Feinnerman and Newman, 1974]) where the model of problem solving is

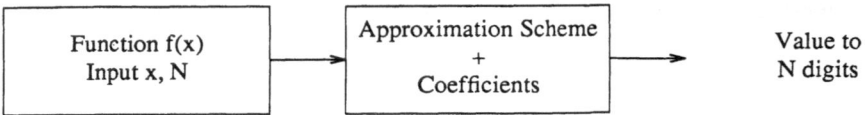

The central question here is: *For a given approximation scheme and class of functions f, find algorithms for the coefficients that give N digits of accuracy with the fewest coefficients.* This framework is less general than that of complexity theory but it is more general than it appears and has a much older and large body of results. Simple examples of this theory are listed below. We use the notation that $C^k$ is the set of functions with $k$ derivatives Lipshitz continuous on an appropriate domain, $N$ is an integer, $f$ is a function, $x$ is a real, polynomials have degree $P$, $L$ is a partial differential operator, $\Omega$ is a domain with boundary $\partial\Omega$.

| Function Class (Problem) | Data | Approximation Scheme (Machine) | Solution Type |
|---|---|---|---|
| $f(x)$ continuous in [0,1] | N | polynomials | real |
| $f(x)$ in $C^k$ | N | polynnomials | real |
| Analytic in unit disk | N | rationals of degree P/P | complex |
| Piecewise in $C^3$ + algebraic singularities | N | cubics with P pieces | real |
| $f(x)$ satisfies $L(f) = 0$ in $\Omega$ $f = 0$ on $\partial\Omega$ | N | quintics with P pieces and in $C^2$ | real |

Observe that the approximation scheme defines the machines. Thus polynomials of degree $P$ means the machine has the program

$$V = c_P$$
$$\text{For } i = P - 1 \text{ to } 0 \text{ do } V = x \times V + c_i$$
$$\text{Value of } f(x) \text{ to } N \text{ digits} = V$$

where the $c_i$, $i = 0$ to $N$, are the coefficients computed by the algorithm of approximation theory. As before, one must be careful in formulating the problem properly so it will not be unsolvable or trivial. The approximation theory framework is more suitable for scientific software because it focuses on functions and reals rather than integers. The small subfield *analytic computational complexity* of computer science is a "merger" of these two frameworks. See, for example, [Traub, 1976] and [Traub and Wozniakowski, 1980].

As an aside, note that there is circularity in mathematics of the definitions of function classes and machines. Those functions which are $C^k$ in [0,1] are exactly those functions where polynomials achieve accuracy of $N$ digits with $10^{N/(k+1)}$ coefficients. Similar equivalencies exist for most other standard function classes of mathematics.


## 4. What is a Solution?


The classical solutions of problems are numbers or simple mathematical expressions. As problems have become more complex, we have admitted more complex solutions such as infinite series or products and integral forms. This raises the question of legitimacy for some such solutions are no solution at all. As a concrete example consider the Gamma and Incomplete Gamma functions. Consulting the *Handbook of Mathematical Functions,* National Bureau of Standards, Applied Mathematics Series, Vol. 55 one finds four integral representations, five asymptotic expansions, one continued fraction expansion and one infinite product. All of these are presented as solutions to the problem of evaluating $\Gamma(x)$. However, only the asymptotic expansions based on Stirlings Formula (coupled with the recurrence $\Gamma(x + 1) = x\Gamma(x)$) can be used to obtain actual values of $\Gamma(x)$.

These complex mathematical expressions are really just algorithms for solutions. And, as in the case of $\Gamma(x)$, they might work so slowly as to be useless. Nevertheless, they set the precedent for accepting an algorithm as the solution of a problem. Since algorithms are merely instructions for machines we can, in some cases at least, say that a machine is the solution of a problem. Thus, hand held calculators do solve the problems of arithmetic and elementary functions for "short" numbers.

Many problems are so complex that no exact solution will ever be found in the domain of mathematical expressions, algorithms or computer programs. Thus we must settle for approximate solutions and formalize our thoughts about them. The series

$$e^x = 1 + x + x^2/2 + x^3/6 + \cdots + x^m/m! + \cdots$$

provides a sequence of approximate algorithms for the value of $e^x$ which is quite satisfactory. The best one can hope for most scientific problems is scientific software of the same nature. The software has a parameter or two which, when increased like the $m$ above, produce more accurate results with a reasonable added cost.

To illustrate the difficulties of defining what a solution is, consider the problem of solving partial differential equations (PDEs). We might find solutions which are

*Finite:* For example,

$$u(x,y) = 0 \text{ if } x < y$$
$$= (x - y)^3 \text{ if } x \geq y,$$

*Closed Form:* For example,

$$u(x,y) = e^x + y(\sin x - \cos y)$$

This almost never occurs so we might hope for solutions in

*Symbolic (Analytic) Form:* For example,

$$u(x,y) = \sum_{i=1}^{\infty} \sum_{j=1}^{\infty} \frac{i+j}{(ij)!} x^i \sin(jy)$$

$$u(x,y) = \sum_{i=1}^{\infty} \sum_{j=1}^{\infty} \frac{(1 + \sqrt{i+j})(-1)^{i+j}}{(i+j) \log (i+2j)} x^i y^j$$

$$u(x,y) = \iint\limits_{R} \sin^2(x - y) f(x,y) + \int\limits_{\partial R} \cos (x)\sinh (y) g(x,y)$$

where $f(x,h)$ and $g(x,y)$ are known data functions. Such solutions might or might not be useful. The first and third examples here probably lead directly to quite effective algorithms for $u(x,y)$, the second expansion is probably useless. The point is that these must be transformed to actual algorithms before one knows the "quality" of the solution.

A more desirable and realistic definition of a solution is to ask for a form as follows.

*Algorithmic Form:*

| *Read* | Ndigits |
|--------|---------|
| *Compute* | K = EFFORT_ESTIMATE ( Ndigits ) |
| *Compute* | $u^e(x,y)$ = SOLVER $(x,y,k)$ |

where EFFORT_ESTIMATE and SOLVER are fixed procedures involving arithmetic and elementary functions (i.e., computer programs) and we know that

$$| u^e(x,y) - u(x,y) | \le 10^{-Ndigits}$$

The infinite series symbolic solutions can usually be recast in this form with little effort.

More specific algorithms for PDEs are

*Finite Difference Algorithm Form:*

| *Read* | Ndigits |
|--------|---------|
| *Compute* | Mesh = GRID_GENERATOR (Ndigits) |
| *Compute* | $u^e(i,j)$ = SOLVER $(i,j,Mesh)$ |

where

$$| u^e(i,j) - u(x_i,y_j) | \le 10^{-Ndigits} \qquad x_i,y_j \text{ in Mesh}$$

Note that this does *not* solve the PDE as expected because a solution is provided only on a certain mesh of points. If that mesh only has two points in it, the solution is surely useless.

*Finite Element Algorithm Form:*

| *Read* | Ndigits |
|--------|---------|
| *Compute* | Elements = ELEMENT_GENERATOR (Ndigits) |
| *Compute* | Coefficients = SOLVER (Elements) |

then

$$u^e(x,y) = \sum_{i=1}^{k} (Coefficients\_i)\phi_i(x,y)$$

where the sum is over all $k$ elements and the $\phi_i$ are basis functions such as cubic polynomial patches and where

$$| u^e(x,y) \_ u(x,y) | \leq 10^{-Ndigits}$$

Since finite element methods produce solutions defined everywhere, we do not have the difficulty seen for finite difference methods.

Pursuing this further in the case of PDEs, we see that an algorithm for solving a PDE should produce a second algorithm which evaluates the approximate solution. The desirable properties of the evaluation algorithm, call it UVAL, are as follows: Given $(x,y)$ then $| UVAL(x,y) - u(x,y) | \leq 10^{-Ndigits}$ and UVAL is evaluated with constant effort (i.e., independent of $x$ and $y$). This view allows a finite difference method to solve a PDE provided one is also provided with an interpolation procedure which preserves the accuracy of the table of approximate values. It may be difficult to find such an interpolation procedure.

We conclude from this discussion that, for many scientific problems, a mathematical analysis of the best ways to solve the problem requires a serious study of what a solution is.

## 5. What Should a Mathematical System Be?

At the conference there was a panel consisting of Carl de Boor, Bradley Lucier, Richard McGeehee and Clarence Lehman which addressed this question and generated lengthy discussion from the audience. It was generally agreed that is is practical today to build a computer system (both hardware and software) that

- automates the algorithmic aspects of undergraduate mathematics,
- communicates in standard mathematical notations and terminology,
- provides high resolution graphics support,
- costs much less than the equipment typically found in a scientist's lab.

While there was not agreement on the details of cost and performance, it is highly likely that for a few thousand dollars one could obtain high reliability in performing the algorithms, very substantial computing power for them and a much higher "thinking level" for computations and problem solving.

Most of the discussion involved opinions as to why such a system does not exist. It was agreed that it is obviously highly desirable to have such systems. There have serious research efforts on mathematical systems since the middle 1960's (see, [Klerer and Rheinfelds, 1968] for an early survey) and one might conclude that it is not yet known how to build them. This is not the case. One opinion is that the research groups spend most of their effort in adding sophisticated facilities rather than in building practical, usable systems. Several members of such groups indicated that they were, in fact, operating this way. This phenomenon has earlier been discussed by [Rice, 1973].

A more likely reason for the lack of such systems was advanced. Classify software into three groups as follows: 1) Basic tools that are essential to get anything at all done with reasonable effort. This includes language compilers, graphics packages, file systems, mail systems and so forth. 2) Problem solving tools that directly solve users' problems. This includes income tax packages, structural engineering systems, transactions systems for banks and so forth. 3) Generic tools that are widely applicable but usually not solvers of the final problems. This includes compiler generators, statistical libraries and mathematical systems. It is much harder to finance the generic software tools. Manufacturers of equipment must produce basic tools and end users are willing to pay good prices for problem solving tools but almost no one is willing to finance generic tools. This is quite reasonable in most cases, no one person or even group gets enough benefit themselves from these tools to afford the high development costs. A company that builds and markets generic tools finds that users will pay much less for them than for problem solving tools (which may be much cheaper to produce). Thus, no organization gets enough benefit from a mathematical system to justify its cost and software companies do not see users willing to pay enough to justify their risk in building and marketing it.

Other considerations were advanced and it is probable that lack of a good mathematical system is due to a complex combination of reasons. The difficulty is not with deciding what such a system should be but rather locating a mechanism to finance its development.

## 6. Symbolic Computing

Symbolic computing has established itself as a small, active subfield of computer science. However, it seems to me that it has not flourished as it should. A great deal of what mathematicians do is symbolic and symbolic computing has the potential for great impact on mathematics. Thus mathematicians should be concerned about why symbolic computing has not flourished. Before considering this, I make a few remarks about the field.

Symbolic computing has three major branches. The first branch is college algebra, calculus and applied mathematics. This is most relevant for scientific software as there are hordes who need to differentiate, integrate, expand in series, change coordinates, manipulate expressions, and so forth. The second branch is in abstract algebra where the operations of rings, fields, groups, etc., are involved. This branch is one that may eventually have the largest impact on mathematics. The third is in logic computations. In fact, I personally do not view this as part of symbolic computing. Its current inclusion is, I believe, due to people classifying everything non-numeric as symbolic. It is more reasonable to think of the subfields of computing as numeric (analysis), symbolic (algebra), logic, geometric (geometry/topology) and perhaps others.

I see four possible reasons for symbolic computing not to have matured as I expected. First is the lack of demand. There is less application of symbolic methods than numeric, but I do not think there is so much less. Great synergy can take place between numerical and symbolic computing and I think the time will come when people will not understand how they becomes so separated. Thus I do not consider the lack of demand as a primary reason.

Second it the lack of adequate computing power. This was a very significant problem in the 1960's and 1970's when memory was very expensive. Most computers simply had too little memeory for serious symbolic computing. This situation has changed dramatically as we now see ordinary computers with 8, 12, 24 or more megabytes of memory, even simple workstations may have 4 megabytes, and personal computers with 1 or 2 megabytes of memory will be commonplace soon. So, while lack of computing power is no longer a barrier for the development of symbolic computing, it was for a 12 to 15 year period beginning in the early 1960's.

Third is the lack of appropriate languages or the incompatibility of symbolic processing with Fortran. It is true that Fortran must be substantially modified to allow for symbolic operations and it is plausible that some new language would be used instead of modifying Fortran. However, I believe that there is no magic in Lisp or even in functional languages in general. Serious symbolic computing has been done in completely Fortran environments and, more to the point, it is not so difficult to mix Lisp and Fortran in a disciplined way. This allows for symbolic computing to flourish in the midst of a numerical computing environment. Thus I do not consider language issues to be a primary reason.

Fourth is lact of appropriate computers systems. It is true that numerical computing has been Fortran based and symbolic computing has been Lisp based. And there have been computers built with special hardware to support Lisp. So it might be that the lack of Lisp machines has kept symbolic computing from flourishing. I do not consider the differences to be of primary significance and symbolic computing could have flourished without special hardware.

I conclude that there were unavoidable reasons for symbolic computing to develop slowly (primarily inadequate computer memories). Those reasons are now gone and mathematics can hope for rapid progress in this area.

## 7. Algorithms for Geometry

The algorithms of geometry are surprisingly complex. People look at pictures and do things (see patterns, intersections, move things) easily that are in fact very complex. Some ot these things cannot, as yet, be computed reliably. My message is simple: *the algorithms for geometry are much harder than one expects.*

The situation is further complicated because *there still is not a satisfactory way to represent general three dimensional objects for computation.* The objectives for representations are threefold:

- *Simplicity*: An object which is fairly simple should have a fairly simple representation.
- *Manipulation*: It should be easy (or at least feasible) to display objects, intersect objects, identify components (boundaries, holes, etc.) of objects or move them.
- *Generality*: One can accurately represent common objects and preserving important properties like smoothness or convexity.

We examine briefly the three principal representation techniques.

The *parametric representation* is two varieties: *explicit*, for example,

$$x = f_1(u,v,w), \quad y = f_2(u,v,w), \quad z = f_3(u,v,w)$$

$u$, $v$, $w$ in unit cube

and *boundary*, for example,

$$\text{side 1:} \quad x = f_1(u,v), \quad y = f_2(u,v), \quad z = f_3(u,v)$$

$u$, $v$ in unit square

and the object is the interior of the region with boundaries side 1, side 2, side 3,etc. It can be a substantial effort to obtain a parametric representation (suppose the object already exists). The objectives of generality and simplicity are reasonably met but certain manipulations are computationally expensive. Examples of this include a) determining if a point is inside or outside, b) making intersections or contacts with other objects, and c) checking to see if the sides match properly to define an object.

The *functional* (algebraic or explicit) *representation* is to define the object as the set $(x,y,z)$ so that $f(x,y,z) \geq 0$. It can be even harder to obtain a functional representation of an existing object than a parametric one. Some manipulations are easy (e.g., determining if a point is inside or outside) and there is no need to check the matching of sides. It is not clear how well the objectives of simplicity or generality are met.

The *constructive solid geometry representation* or building blocks is to take a small set of generic objects (e.g., spheres, cylinders, planes, parallel grids) and construct objects by unions, differences and intersections. For example (using − for difference and + for union) we might have object = cube – (vertical cylinder) + spherical cap on east face

These representations are the easiest to manipulate and in many applications they are both simple and general. Their principle weakness is generality. It is difficult, for example, to represent a free form such as a face, a turbine blade or a car body smoothly, with modest accuracy and appropriate simplicity.

Since none of these representations is completely satisfactory, some applications use more than one which introduces the problems associated with transforming one representation to another. These transformations may be both computationally expensive and mathematically difficult to determine.

To illustrate the surprising complexity of algorithms for geometry, consider Figure 1 where a simple domain with four bounding sides is shown along with a rectangular grid. I have written a Fortran program to compute the points of intersection of the grid with the boundaries of the domain. This algorithm is very complex even though a person can identify the points very quickly. To quantify this, the code is over 2000 lines long which is more than enough to write many high performance, sophisticated PDE solvers on rectangular domains. Furthermore, the PDE solver will be more robust, this geometry code has been tested very extensively and yet cases of unsatisfactory performance still arise.

Not only are algorithms for geometry complicated, they are intrinsically hard. For example, consider the following two related problems:

1) Find a path for an object through an obstacle course.

2) Find the shortest such path.

We ''solve'' such problems everyday as we walk across campus, a dance floor or a parking lot. We feel these are easy tasks and even though we do not get the absolutely shortest path, we feel we come close. We are wrong, these are not easy tasks. Mathematical arguments of complexity show that such problems are intrinsically hard and experience at studying such problems confirms that there are no easy, quick ways to find shortest or almost shortest paths.

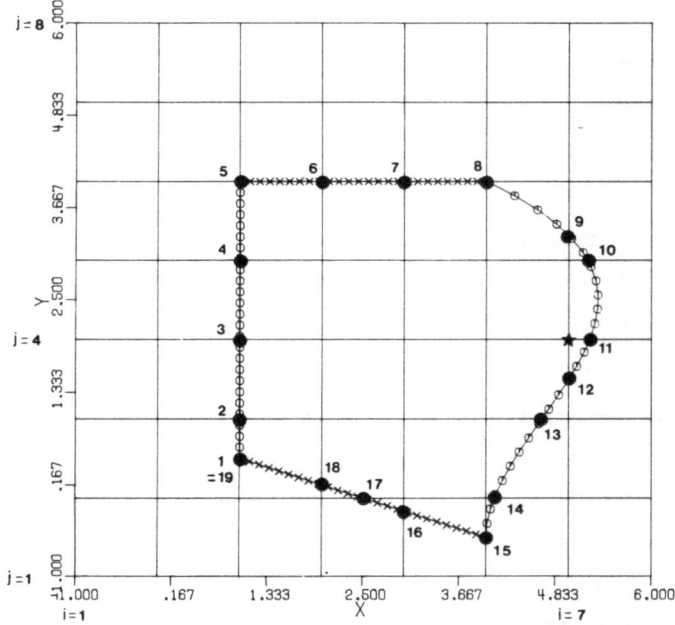

Figure 1. A domain showing the intersections of its four sides with a rectangular grid.

For another example, consider another pair of related problems:

1)   Smooth off the sharp edges of an object nicely.

2)   Blend 2 or 3 surfaces together smoothly where they join.

Again, in practice, people just "do" these things as they make or design objects. The Computer-Aided Design and Computer Aided Manufacturing (CAD/CAM) community has been challenged by this problem for about 30 years, ever since they started to automate some steps of design and manufacturing. Although algorithms have been found which work well in many applications, they are not robust and reliable. People are still trying to understand exactly what we mean when we say "smooth" or "blend" and how one can achieve this. The most recent advances on these problems have been by applying algebraic geometry techniques.

## 8. Geometric Computing

One of the exciting frontiers just opening up is geometric computing. The previous section discussed algorithms for geometry but these are things embedded in numeric or symbolic computing. We will see computations that directly manipulate geometric objects and geometric systems will include a wide variety of geometric and topological operators. One will not manipulate objects by applying operations (tediously detailed) to concrete representations but rather one will do manipulations explicitly in a geometric language.

There is no existing mature language for geometry beyond the colloquial one we use in everyday life. It is not at all obvious how the language will appear, but it must have:

> *Variables (objects)*:
> lines, circles, boxes, spheres, ...
> curves, surfaces, boundaries, tangents, corners, ...
> regions, objects, pieces, interiors, ...

> *Attributes*:
> straight, smooth, convex, closed, ...

> *Operators*:
> rotate, move to, bend, join, add (union), ...
> smooth, stretch, shrink, map, ...

I give an example geometric algorithm for a PDE problem.

> Display domain *A*
> Expand corner 3 to a side
> Map results onto a rectangle
> Do a boundary layer map toward side 2
> Shrink domain locally toward point *x*
> Overlay the domain with a rectangular grid
> Transform the PDE on *A* to the current domain
> Solve resulting PDE using spline collocation
> Plot PDE solution on domain *A*

This algorithm is from real computations implemented in a numeric PDE oriented language. The code for that computation was much longer and less clear. To express this algorithm in a standard language like Fortran or Pascal would require thousands and thousands of lines of code.

Geometric computing requires even more computer power than symbolic computing. An adequate workstation needs perhaps 20 MIPS (million instructions per second), 5 MFLOPS (million floating point operations per second) of processing power (about 20 to 50 times a standard VAX 11/780 computer), color graphics displays with high resolution and sophisticated built-in graphics processing, 10-40 megabytes of main memory plus 1-5 gigabytes of auxiliary memories (disks). There will be millions of lines of Fortran or C or similar code (if such languages are used for the implementation). Hardware with these characteristics will cost perhaps $100,000 in 1988-89. Five to seven years later we can hope for such machines to cost 10 or 20 percent of this. The software will cost enormous sums and is unlikely to appear in 5 to 7 years.

Geometric techniques have always been one of the most powerful approaches to problem solving. There is a huge body of knowledge about geometry and topology but I doubt that is is well organized for the task of creating a system for geometric computing. Once this task gets under way, a host of problems will arise to challenge mathematicians and computer scientists. And, quantum leaps will be made in our power to solve problems.

## 9. Mapping Problems onto Machines

Our earlier discussion of problem solving centered on finding good or best algorithms for particular machines. These were rather simple abstract machines but, until recently, they modeled well the computers in actual use. The advent in the 1980's of complex machines using several – or hundreds of – processors in parallel has introduced a new and essentially difficulty into problem solving: *how do you map a problem's structure onto a machine to take advantage of the machine's power.* Worse that losing the simplicity of the old von Neumann architecture is that we must face dozens of different architectures, some radically different from others.

If we had enough time and money, we would take each pair of problem and machine and then analyze how to best create algorithms on this machine for this problem. This approach is used only for the most important problems and the related mathematical problems are similar to those of the slightly different approach presented next.

Realistically, we have only a few choices of reasonably efficient algorithms for a given problem. The practical approach is to take each of these algorithms and structure (or transform) them to a form that is quite flexible and amenable to mapping onto various machines. We call this a *computational structure* and it consists of:

- a precedence graph whose nodes are computational modules. These modules might range from a single or handful of instructions (as in microcoding of processors) to a collection of dozens of lengthy procedures (as in some scientific applications).
- information at each node of the graph on the processing time, memory and data access required by the modules,
- information at each arc joining nodes about the amount of data that must pass between the nodes.

A simple example of a computational structure is given in Figure 2. Figure 3 shows a realistic one with only part of the information given about a partial differential equation's (PDE) solver. See [Houstis, Houstis, Rice, 1987] for more details about this example.

One also has a *machine structure* which is a collection of processors and memories connected by a communication network. The traditional von Neumann machines have one processor, one memory and no communication network, the simplest possible machine structure. Figure 4 shows the schematic of a hypothetical, but currently plausible, machine structure. This machine has 31 ordinary processors (P) with 1 megabyte of local memory, 4 faster processors (P*) with 2 megabytes of memory, 4 vector processors (V), 14 modules with 4 megabytes global memory, 8 disks and 25 modules with 128 kilobytes of fast global memory. It also has a complicated communication structure.

The *mapping problem* may now be stated. Given computational and machine structures, determine an assignment of the computational nodes to processors, and memory and data nodes to memories so the whole computation is completed as fast as possible. It is well known that such scheduling or assignment problem are exponentially hard to solve, i.e., it is not feasible to find optimal solutions.

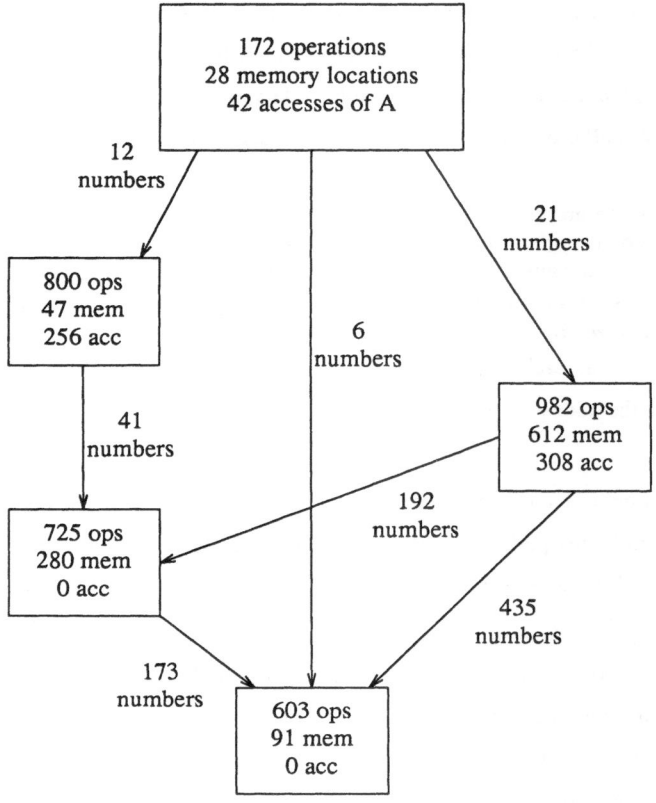

Figure 2. Simple example of a computational structure. Each box contains the number of operations (ops), memory locations (mem) and accesses made to the data labeled *A* (acc). Each arc is labeled with the data that passes between nodes.

There are two general approaches to solving this problem. First is to restrict the computational and/or machine structures to some simple class. Then optimal or nearly optimal mappings can be obtained by mathematical analysis. For example, one might assume that the computational structure is a regular tree and that the machine is a rectangular mesh of identical processors and memories. Second is to apply heuristic or approximate mapping algorithms. This can be done by the programmer while he is writing the program, by the compiler as the code is translated into machine language, by the loader as the pieces of the computational structure are organized to be placed in the machine, and during execution. In fact, sometimes one can apply all four of these

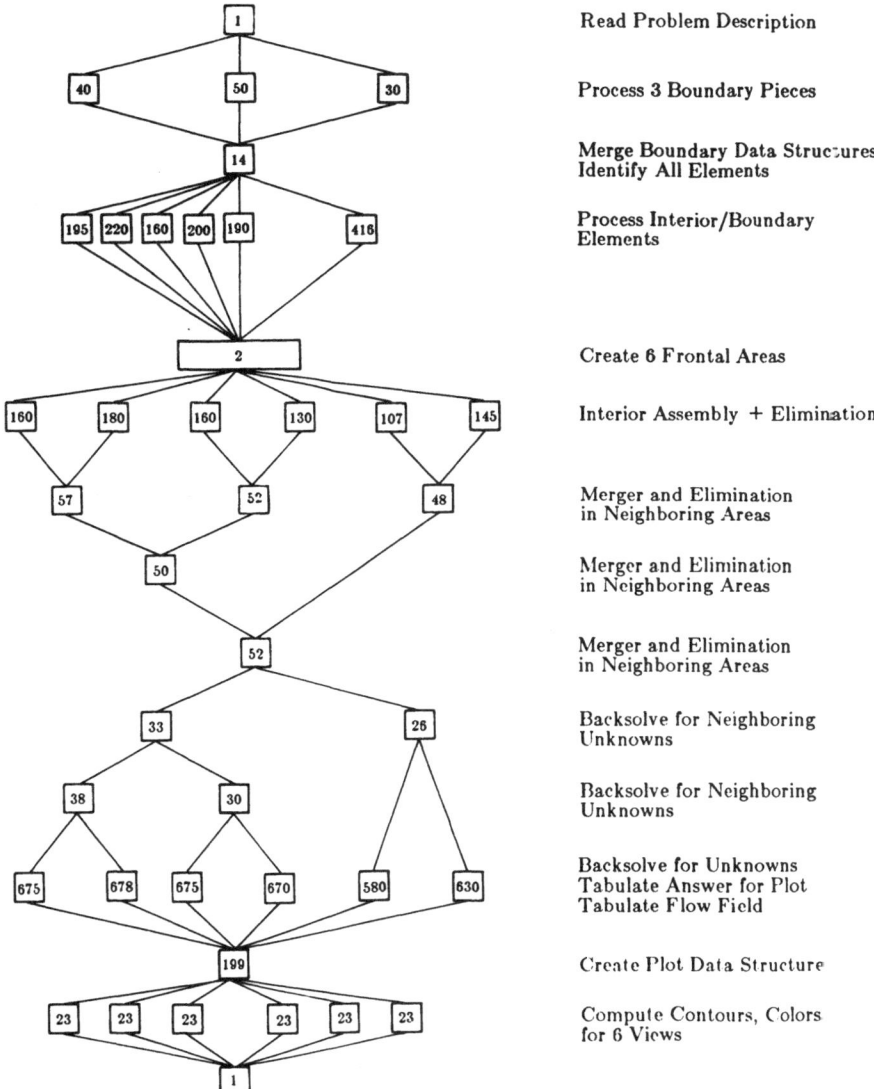

Figure 3. Partial computational structure of a PDE solver. The numbers at the nodes are the thousands of arithmetic operations to be performed there. The memory, data access and data passed between nodes is not shown.

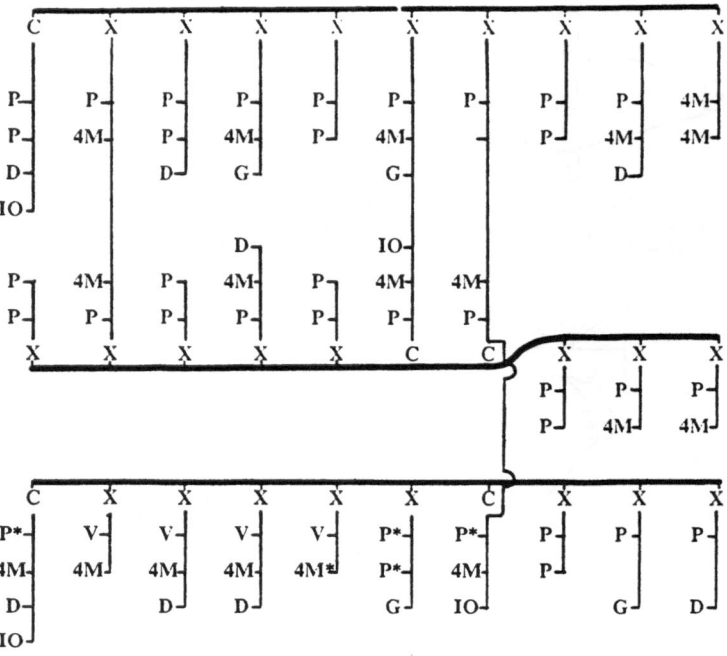

Figure 4. Schematic of a hypothetical machine with a complex architecture.

techniques to the same computation. Results from the first approach can be incorporated into the second to determine some pieces of the mapping. See the paper [Berman, 1988] in this volume for more discussion of the mapping problem.

Figure 5 shows the result of a heuristic method applied to the Cholesky factorization of symmetric matrices. The algorithm has been partitioned to be suitable for a machine with five processors. In Figure 5 the nodes are only numbered and the amount of data transferred between nodes is given. The machine has all identical processors and a simple bus communication network. The heavy lines show the nodes assigned to the processors.

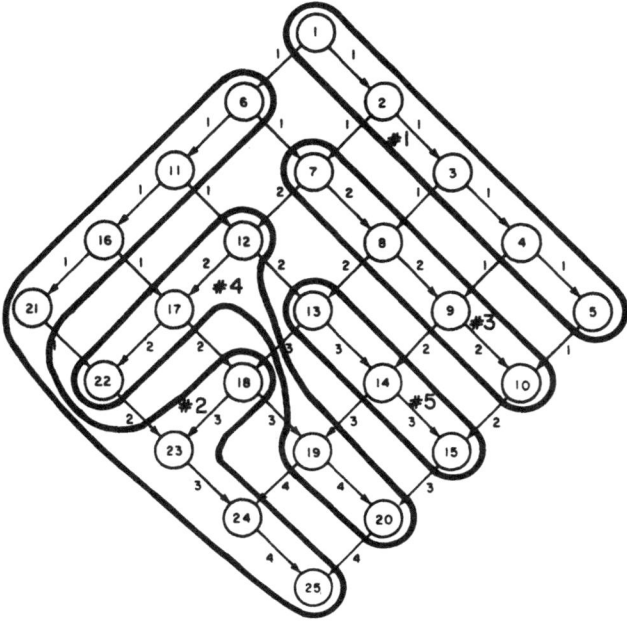

Figure 5. Computational structure of a parallel Cholesky factorization algorithm showing an assignment of the nodes to a machine with five processors.

## 10. Smart Algorithms or Adaptive Methods

The idea of having a numerical computation adapt itself to the problem at hand appeared in the early 1960's in a) ordinary differential equations programs where variable step sizes were chosen using local error estimators, and in b) numerical integration algorithms where intervals to be subdivided were chosen again using local error estimators. Before that computer programs used a fixed algorithm for all members of a problem class. Of course, when computing was done by hand (or desk calculator) adaptive methods were the norm. The human computer observed every step of the computation and adjusted many things to make them more efficient. These techniques were not unified into formal algorithms as required for computing machines.

Adaptive algorithms have led to dramatic advances both in theory and practice. For example, classical numerical integration of a function $f(x)$ with a singularity at $x_o$, say $f(x) = (\sqrt{x} + 3x^2)(\cos(x^2 - x^3)^2/(1 + 3\sqrt{x^2 + 1})$, requires one to locate the singularity, determine its nature and use a special technique (e.g., Gauss quadrature with weight function, change of variable, series expansions, etc.). A modern adaptive integration algorithm can achieve good efficiency without knowing (as input) either the location or nature of the singularity. It determines both approximately during the computation and, as the accuracy requested increases, it achieves more accurate approximations. At the end of this section we sketch a simple, educational adaptive integration algorithm.

The idea of adaption has spread throughout numerical computation. For example, in solving partial differential equations one now sees time varying discretization meshs and different order discretizations being determined by adaptive criteria. Both these techniques come from the early 1970's algorithms for ordinary differential equations. These algorithms permitted, for the first time, the reliable and efficient solution of general initial value problems in ordinary differential equations. Similarly, these techniques are used in the amazingly robust and efficient algorithm for numerical integrations that were develop in the late 1970's and early 1980's. Of course, the application of these techniques in more than one dimension considerably complicates both the algorithms and their analysis.

Closely related is the concept of *polyalgorithm*, introduced in the middle 1960's in an attempt to achieve both efficiency and reliability. The idea here is to combine several basic algorithms along with various rules as when to switch from one basic algorithm to another. The rules are based on information gathered by the polyalogirhtm during its computations. For example, in solving $f(x) = 0$ one might start with the secant method which is quite efficient for most functions. But it does require $f(x)$ to be somewhat smooth and if the polyalgorithm senses that the secant method is failing to converge, it can switch to bisection (if two values of $f(x)$ are found with opposite signs) or simple systematic search for small $f(x)$ values. Polyalgorithms are widely used now, especially for optimization and nonlinear equations problems. They are also the precursors of *expert systems* for numerical computation. A big part of what an expert system should do is to choose or change the algorithm used and our experience with polyalgorithms shows that this is both feasible and productive.

These algorithms present mathematics with two challenges, one difficult and one revolutionary. The difficult challenge is simply to be able to analyze specific algorithms as applied to important problem classes. They are inherently more nonlinear and more complex than traditional algorithms and thus we should expect it to take a long time to develop suitable analysis techniques. However, progress is being made and one example

theorem is given in Section 12. That example is typical in that the hypotheses contain smoothness assumptions but these are not reflected in the conclusion. The reason for this is that the very foundation of the analysis has changed.

The revolutionary challenge is to replace the foundation of much of the function theory. Currently it is based on machines that can add, subtract, multiply and divide. That is, polynomials and rational functions are the models for functions. They are inadequate. In spite of the Weierstrass theorem, polynomials are unsuitable models for the functions that occur in the real world. Adaptive methods introduce logical decisions into the system or, in more analytic terms, piecewise rationals. There is overwhelming evidence that piecewise polynomials (or rationals) are very suitable models for real world functions. Once one switches to piecewise rationals (machines that add, subtract, multiply, divide and do logical tests) dramatic changes occur in the associated function theory. For example, the functions six(x), $(x-1)^{3.63}$, $\sqrt{x}$ and $(x-.15)^{1/69.23}$ are all in the same smoothness class. Here smoothness classes are defined (as in current mathematical practice) in terms of the approximation properties of the basic models, piecewise rationals.

**Example: The adaptive trapezoidal rule**

Consider a function $f(x)$ concave on [a,b] and the use of the trapezoidal rule to integrate it. Using five integration points (four intervals), we obtain the situation shown in Figure 6. The areas of the trapezoids below the curve are computed and added to estimate the area under the curve. The areas of the shaded triangles provide error bounds on the numerical integration. See [Rice, 1973] and [Rice, 1983, Section 7.5] for more details. A general framework for adaptive quadrature algorithm is given in [Rice, 1975].

The adaptive strategy now is to subdivide that interval with the largest triangle (the left one is thus subdivided next). This is continued until the error (sum of triangle areas) is as small as desired. This algorithm can be easily carried out with pencil, ruler and paper for a number of steps. One quickly sees that it adapts rapidly to the nature of $f(x)$.

We now test the effectiveness of adaption for the two functions shown in Figure 7. We compare this algorithm with the classical trapezoidal rule, Simpson's rule and a more sophisticated adaptive algorithm CADRE (it is also the program DCADRE used for

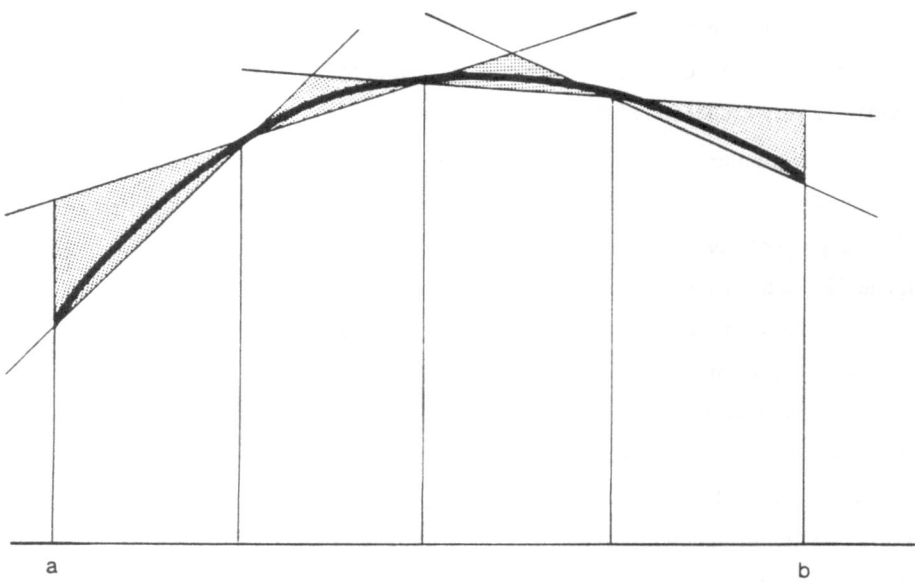

Figure 6. The geometry of the adaptive trapezoidal rule. The curve $y = f(x)$ lies in the shaded triangles whose areas provide error estimates.

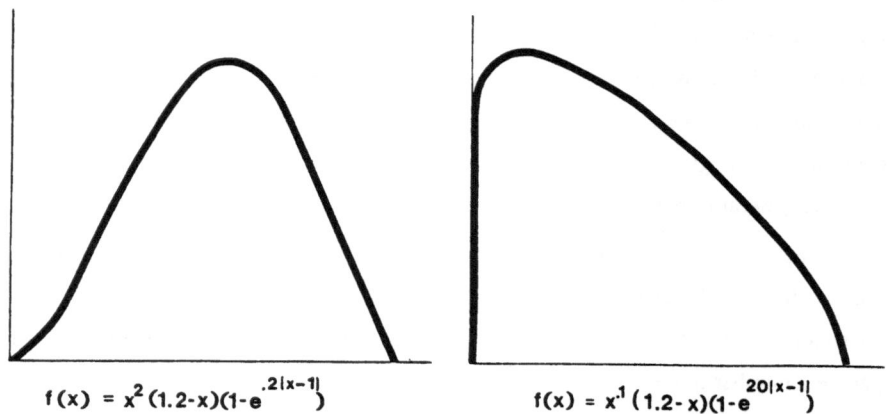

$$f(x) = x^2 (1.2-x)(1-e^{.2|x-1|})$$

$$f(x) = x^1 (1.2-x)(1-e^{20|x-1|})$$

Figure 7. Two test functions for numerical integration, one (left) smooth and one (right) difficult to integrate.

many years in the IMSL library). We compare the algorithms' performance by counting the number of times $f(x)$ is evaluated in order to achieve a given level of accuracy. This ignores the difference in overhead among the algorithms which is not large for any of the algorithms.

Table 1. Results of numerical integration test. The number of $f(x)$ evaluations is given for each algorithm in order to achieve a given level of accuracy.

Smooth case, integral = .009549966

| Method | Accuracy | = | $10^{-6}$ | $10^{-7}$ | $10^{-8}$ |
|---|---|---|---|---|---|
| Trapezoidal rule | | | 68 | 180 | 580 |
| Simpson's rule | | | 13 | 26 | 66 |
| Adaptive trapezoidal rule | | | 22 | 64 | 208 |
| CADRE* | | | 33 | 33 | 41 |

Difficult case, integral = .602297

| Method | Accuracy | = | .01 | .005 | .001 | .0005 | .0001 |
|---|---|---|---|---|---|---|---|
| Trapezoidal rule | | | 36 | 67 | 285 | 510 | 2080 |
| Simpson's rule | | | 23 | 43 | 184 | 340 | 1470 |
| Adaptive trapezoidal rule | | | 14 | 19 | 43 | 60 | 133 |
| CADRE* | | | 33 | 33 | 33 | 33 | 33 |

*CADRE's accuracy is much better than the column heading

The test results in Table 1 show three things: 1) Adaptive algorithms can be very effective in handling difficult problems. 2) Adaptive algorithms perform better even for smooth problems, though not by a dramatic margin. 3) An adaptive algorithm might require more function evaluations for low accuracy results. This is because a good algorithm (such as CADRE) is rather cautious about accepting a result as correct; if we did not know the answer to these problems Simpson's rule would need more than 1470 evaluations to be confident of .001 accuracy for the difficult case.

## 11. Performance Evaluation: How Well Can We Solve Problems?

Since mathematics focuses on problem solving, it must address the question of how well we can solve problems. One issue is the performance of algorithms and this fits naturally into the framework of mathematical analysis. The earlier discussion of complexity theory, analytic complexity and approximation theory shows we have well developed machinery to analyze the performance of computations on abstract machines.

The situation is not nearly so satisfactory for computations on real machines. There are fundamental difficulties in each of the two area where mathematics might and should provide results. The first of these is *performance theorems*. Ideally we want a theorem of the following type: Consider problem class P defined to have properties $p_1, p_2, \cdots$. Then algorithm A always solves problem from class P.

In order for such theorems to be useful, it is essential that the classes P involved correspond well to real classes of problems. That is, mathematics must provide realistic models for real problems and it does not do this well. I estimate that half, at most, of real world problems can be reasonably modeled by currently standard mathematics. Further, a large portion of those problems modeled in applied mathematics do no correspond to any real problems. The difficulty comes primarily from the admissible classes of functions. In mathematics one has classes of functions defined by continuity, derivatives, analyticity, convexity, etc. In the real world one has classes of functions defined by smooth, well-behaved, ramp functions, the geometry of real objects, etc. Real world functions have a finite "scale", $f(x)$ might be smooth at a scale of 1, very rough at a scale of .0001 and randomly defined at a scale less than $10^{-6}$. In mathematics, the "scale" of behavior (or classification) is zero except for properties like convexity. It is not clear whether all real world functions have six derivatives (or are entire) or whether none of them have. It is clear that there is a serious mismatch between models and reality.

A symptom of this difficulty is the problem of *unverifiable hypotheses* which is discussed in the next section. It suffices here to note that if one cannot say whether $f(x)$ has four derivatives or not then a theorem with this as a hypothesis is not useful.

It is unavoidable that performance evaluation must have a large component of experimental work. Our ability to analyze problems, algorithms and machines will always lag behind our ability to formulate problems, devise algorithms and build machines. Perhaps mathematics should not take on a role in experimental science, but it (or perhaps statistics) should provide a proper framework for experimental studies.

One cannot overemphasize the importance of well conceived performance experiments. The traditional approach of consulting experts is unreliable. I believe the following is true (and I have observed concrete instances of this). Pose a simple problem P (e.g., solve 1 equation in 1 unknown in the interval [0, 1]), then consult 10 experts for the best method to use. About 4 to 6 methods will be recommended. If one consults the whole population of experts, almost the whole population of methods will be recommended. Science in general and scientific software is particular cannot advance without reiable information on the relative quality of its techniques.

An important contribution of performance evaluation is in the development of algorithms. In those areas where we have seen considerable advances in algorithms, it is extensive experimentation that has led to some of the key ideas. The paradym is that one creates an algorithm, evaluates its performance (speed or reliability) and then focuses on those instances where performance is poor. With some luck and perseverence, one discovers the ''cause'' of the poor performance and modifies the algorithms to improve it. The eventually best algorithms are discovered through systematic performance evaluation and were unknown beforehand.

Figures 8 and 9 show two simple examples of performance evaluation for algorithms to solve the Poisson problem on the unit square. In Figure 8, we have discretized the problem using standard, simple finite differences and then solved the resulting system of linear equations. We have plotted (for a particular problem) the accuracy achieved versus the computer time used for seven methods (see, [Rice and Boisvert, 1985] for specific definitions). We see that Gauss elimination (BAND GE) is the worst method, it is the algorithm that existed before 1945. Over the past 30 years we have discovered various iteration algorithms (represented by JACOBI CG here), fast Fourier transform algorithms (represented by FISHPACK-HELMHOLTZ and FFT-9 POINT) multigrid (represented by MULTIGRID-MG00) and others. The results of 30 years and perhaps 1000 papers is a speed up of about two orders of magnitude (a factor of 100) in solving this problem.

Figure 8 shows the results of using different discretizations for the same problem. This happens to be a rather easy problem, so high accuracy is obtained quickly compared to many realistic problems. The best result, multigrid, for ordinary finite differences is shown and we see that it is the worst algorithm. Speedups of five orders of magnitude are possible by using better discretizations. I conclude that the primary focus of research effort should be on discretization rather than solving linear systems. This conclusion was reached by a number of people about 10 years ago and they began advocating it. Even today, now that the experimental and theoretical evidence is overwhelming, this conclusion is still only accepted by a small fraction of those solving partial differential equations. The reason for this is that the scientific software field has not matured enough to place scientific experimentations (performance evaluation) in its proper role.

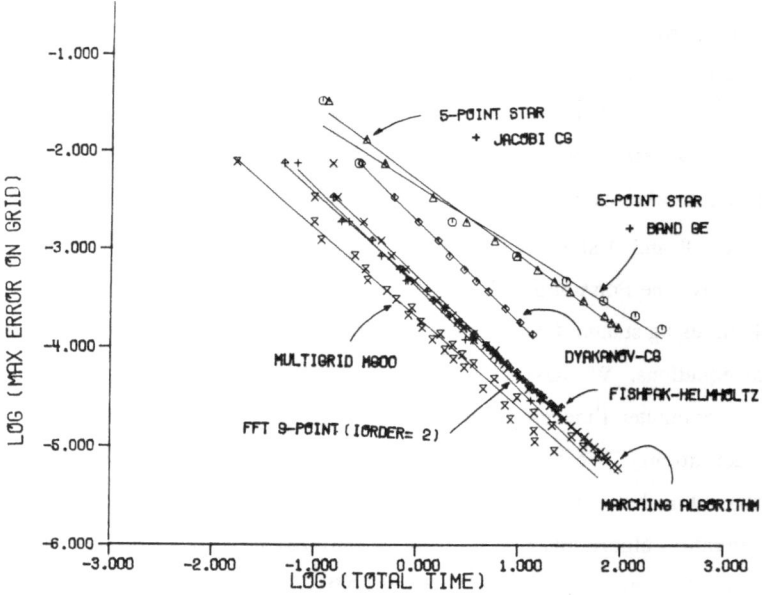

Figure 8. The performance ( accuracy versus machine time in seconds) of seven methods to solve the linear system generated from discretizing $u_{xx} + u_{yy} = f(x,y)$ by standard finite differences.

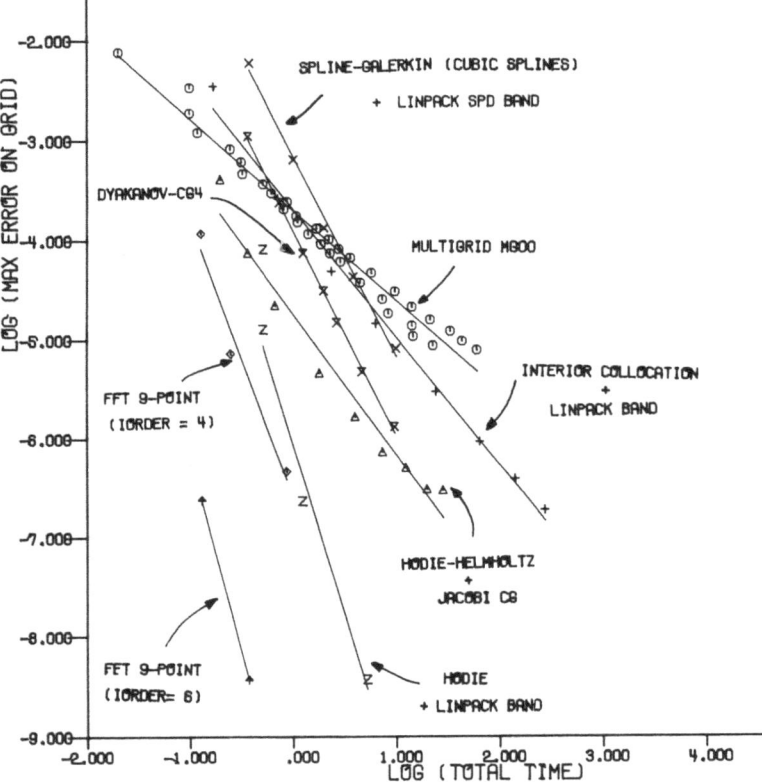

Figure 9. The performance (accuracy versus machine time in seconds) of eight discretizations plus linear equation algorithms to solve $u_{xx} + u_{yy} = f(x,y)$.

Once one accepts experimental performance evaluation as an essential part of scientific software, one faces another barrier. The approach is to state, for example:

*Hypothesis*:    Algorithm A is more efficient than B is integrating functions on [0 , 1].

One then carries out an experiment by creating a population P of functions on [0 , 1] and performing the

*Experiment*:    Randomly sample the population P, measure efficiency and draw conclusions, if any, from statistical test S.

The barrier arises at selecting the population P. If we select a mathematical population (e.g., functions with 3 continuous derivatives) then, with probability 1, members of this class are irrelevant to real world problems (e.g., their fourth derivatives have infinitely many discontinuities). This is just another symptom of the inadequacy of mathematics to model real functions. If we collect a thousand or a million functions from all kinds of real world applications, then we avoid some objections to the experiment but introduce others. For example, how do we know our collection process was not biased in some fundamental way?

We need statistical tests which apply to incompletely known populations. This is almost a contradiction in statistical terms, but this need is not restricted to the performance of scientific software. As a simpler example, assume we wished to show that DDT kills insects. It is estimated that only 1/3 of all insect species have been identified, so we have considerable difficulty here. Even if we showed that DDT kills all known species, the probability that DDT kills a given kind of insect might be only 1/3. Most people would find such a weak conclusion difficult to consider seriously.

## 12. Verifiable Hypotheses: Does Mathematics Model Reality?

The preceding two sections have raised the issue of whether traditional function theory provides an adequate model for real problem solving. The presentation suggested strongly that the traditional mathematical model is inadequate. Functions classified as "smooth" or "well behaved" by function theory may be unrealistic and pathological when viewed by a scientist while many functions he sees as "smooth" or "well behaved" are not so viewed by function theory. Concrete evidence of this mismatch occurs when one finds that a modestly accurate approximation (say good to 3-4 digits) to a simple curve (simple as seen by a scientist) requires a polynomial with degree in the hundreds or even thousands. Such polynomials are almost always intractable to compute or use and, perhaps worse, they have wildly oscillating derivatives where as the curve (or its underlying function) does not. Such curves occur in nature very commonly.

This issue appears again when we attempt to prove results about scientific algorithms. The typical theorem has hypotheses like

*Let f (x) have four continuous derivatives.....*
*Let f (x) have m + 1 derivatives with.....*
*Assume f ′′′(x) is bounded by k .....*

Such hypotheses cannot be verified in practice and the resulting theorems may have little direct value.

There are directly conflicting plausible arguments about the "true" nature of real world functions. First, one can argue they are infinitely smooth except for a finite (small) number of discontinuies that correspond to discrete events like turning off a switch or changing composition of material. This argument leads directly to accepting piecewise rational functions as the appropriate mathematical models.

Second, we can argue that the real world is inherently discontinuous everywhere, its "microscopic" structure is either discrete or random or both. In any case, the mathematical definitions of continuity, derivation, etc., do not apply because, at some fine scale of examination, the functions are undefined or discrete or something intractible. The implication of this view is that the concepts of smoothness and behaviors of functions are related to a scale and that an adequate mathematical model must take this into account.

Third, one can argue that all real world functions are continuous and, hence, all their derivatives are also. This is supported by the proof that all *computable functions* are continuous, a standard result from theoretical computer science. The computable functions are, of course, the only functions that occur in scientific software and it is hard to accept, at least intuitively, that there are real world functions which are not computable. On the other hand, the function sign (x) looks very computable and discontinuous. This theorem stands in direct contradiction to the first intuitive view and it makes much of modern mathematics irrelevant. This fact underscores that we do not yet understand the true nature of the relation between computation, mathematics and the real world.

I cannot resolve these contradictions but the fact that mathematical function classes fail to adequately model reality is leading to more and more difficulty as we attempt to make better analyses of problem solving and scientific computations. As an example, I point out that a number of theorems have been published which "prove" that adaptive computation methods work no better than non-adaptive ones. Such nonsense stems directly from the shortcommings of mathematics in modeling real world problem solving.

As an example of a small step toward verifiable hypotheses, I state a theorem on the convergence of the adaptive trapezoidal rule briefly presented in the preceding section.

Let

$$\text{If} \ = \int_a^b f(x)\,dx$$

$$Q_N f \ = \ \text{Quadrature result using } N \text{ values of } f(x)$$

Assume that

a)  $f(x)$ has $p$ continuous derivatives except for a finite number of algebraic singularities $s_i, i = 1$ to $k$. Further

$$|f^{(p)}(x)| \leq K \prod_{i=1}^{k} (x - s_i)^{\alpha - p}$$

holds for some constant $K$ and $\alpha > -1$ .

b)  we know a characteristic length (scale) $\lambda(f)$ so that local error estimates are valid on intervals of length less than $\lambda(f)$.

*Theorem [Rice, 1975]. There are adaptive quadrature algorithms so that, for some constant $C$,*

$$|\text{If} \ - Q_N f| \leq C/N^p$$

*Corollary:* Let $P = 2$ and $\lambda(f) =$ one fifth the minimum distance between inflection points, cusps, the $s_i$, $a$ and $b$. Then the theorem holds for the adaptive trapezoidal rule algorithm.

The unusual feature of this theorem is that it makes a direct scale assumption. If one views all functions as piecewise smooth (as many do), then the hypotheses make explicit that one must identify all the discontinuities and then one can proceed with confidence to a computation. The scale $\lambda(f)$ is used by the algorithm and a more satisfactory result would show explicitly how the constant $C$ depends on $\lambda(f)$ and $p$.

# REFERENCES

Aho, A.V., J.E. Hopcroft and J.D. Ullman, *The Design and Analysis of Computer Algorithms*, Addison Wesley (1974).

Berman, F., The mapping problem in parallel computation, this volume, (1988).

Davis, P.J., Fidelity in mathematical discourse: Is one plus one really two?, *Amer. Math. Monthly*, 79(1972), 252-263.

Feinerman, R.P. and D.T. Newman, *Polynomial Approximation*, Wilkins and Wilkins (1974).

Houstis, C.E., E.N. Houstis and J.R. Rice, Partitioning PDE computations: Methods and performance evaluation, *J. Parallel Computing*, 5 (1987), 141-163.

Klerer, M. and J. Reinfelds, *Interactive Systems for Experimental Applied Mathematics*, Academic Press (1968).

Rice, J.R., NAPSS-like systems: problems and prospects, *Proc. Nat. Computer Conf.* (1973), 43-47.

Rice, J.R., An educational adaptive quadrature algorithm, *SIGNUM Newsletter*, 8 (April 1973)

Rice, J.R., A metalgorithm for adaptive quadrature, *J. Assoc. Comp. Mach.*, 22 (1975), 61-82.

Rice, J.R., *Numerical Methods, Software and Analysis*, McGraw Hill (1983).

Traub, J.F. and H. Wozniakowski, *A General Theory of Optimal Algorithms*, Academic Press (1980).

Traub, J.F., *Analytic Computational Complexity*, Academic Press (1976).

# THE MAPPING PROBLEM IN PARALLEL COMPUTATION*

FRANCINE BERMAN**

**Abstract.** The *mapping problem* is the process of implementing a computational task on a target architecture in order to maximize some performance metric. This problem is fundamental to parallel computation and researchers have taken a number of different approaches to solve it. This paper surveys and illustrates the characteristics-based approach, language-based approach, linear algebra-based approach and graph-based approach in current research on the mapping problem. Some general themes of these approaches are discussed and contrasted in the last section.

## Section 1. Introduction

With the advent of VLSI technology and the growing need for faster and more powerful machines, parallel computers have become the focus of an increasing effort in research and development. The goal of this effort, that of speeding computation, is being accomplished by the investigation, design and development of parallel machines, algorithms and tools. This research effort becomes even more timely as society becomes increasingly dependent on computers, and the existing and proposed machines employ technologies near the limits of their capabilities.

Multiprocessors are being used by a growing number of researchers and practitioners. To program these machines (efficiently or not), the multiprocessor user must generally know a number of low-level details about the machine such as the number of processors, memory organization, communication and synchronization protocols, etc. In particular, a user who wishes to implement a high-level parallel or sequential algorithm must modify the program extensively to run it on most existing parallel machines. This contrasts dramatically to the sequential environment where an algorithm may be represented in terms of any one of a number of high-level languages, and existing (and efficient) systems software performs the necessary lower-level tasks such as compiling, processor allocation, time-sharing, load-balancing, etc. The lack of similar tools in the parallel environment contributes to the difficulty of programming multiprocessors, and renders them less accessible to the general user community.

The implementation of algorithms on parallel machines is loosely defined as the *mapping problem*. The term "mapping problem" has been used to refer to the allocation

---

* This research supported by the Office of Naval Research Contract No. N00014-86-K-0218.

** Department of Computer Science, University of Calif., San Diego, La Jolla, CA 92093

of processes of a particular set of parallel algorithms on a given multiprocessor architecture, the compilation of sequential programs to an intermediate form executable on a multiprocessor, the scheduling in time of computations on systolic array machines, etc. Clearly, an instance of the mapping problem depends upon both the specification of the algorithm domain and the underlying multiprocessor architecture(s). Moreover, the level of abstraction of these specifications determines the format of appropriate mappings and guides the designation of cost and performance measures for their evaluation. The mapping problem is fundamental to parallel computation. Solutions to the mapping problem both enhance our understanding of parallel algorithms and architectures and provide tools which enable multiprocessor users to program in more high-level environments.

In this paper, we provide a general (although not exhaustive) overview of research on the mapping problem. A number of interesting approaches have been used to investigate the mapping problem and we describe the main directions and give examples of current research in Section 2. We summarize and contrast these approaches in Section 3. Note that the work described in this paper is only a small part of a large body of research on the mapping problem, and we apologize to researchers excluded from this article because of space constraints.

### Section 2: Current research on the mapping problem

As stated in the last section, the mapping problem is a loose term which can be rigorously defined and evaluated only when the algorithm and multiprocessor models have been specified. A number of different approaches have been taken in specifying these models, each of which has focused on a different set of algorithm features and multiprocessor characteristics. These approaches were taxonomized at a workshop held in Fall 1986 sponsored by the National Science Foundation and Carnegie-Mellon University [30]. The focus of this workshop was to assess the status of research in areas related to performance-efficient parallel programming and to discuss productive future directions. During these discussions, the following basic approaches were identified with respect to the mapping problem: the characteristics-based approach, the language-based approach, the linear algebra-based approach and the graph-based approach.

Each of these approaches presupposes a particular specification of the mapping problem. The *characteristics-based approach* assumes the broadest interpretation for the algorithm and multiprocessor domains but both must be identifiable by a given set of features or characteristics. The *language-based approach* presupposes that the algorithm has been given in textual form and the mapping process takes the form of a compilation process into a low-level intermediate form (typically with explicit parallelism) executable on a multiprocessor. The *linear algebra-based approach* assumes that the algorithm can be represented within a matrix format and that the mapping takes the form of a series of transformations from the algorithm matrix into the multiprocessor matrix (usually some sort of systolic array). The *graph-based approach* typically assumes that the processes of the algorithm can be identified in some way with a communication graph and additionally that the multiprocessor can be represented as a processor interconnection network. The mapping process then becomes a graph partitioning and/or embedding problem. Each of these specifications describes an instance of the mapping problem. In this section, we discuss and illustrate current work in each of these research directions.

**The characteristics-based approach**

The characteristics-based approach has primarily been developed by Jamieson ([18], [13]). The basic idea is to represent the mapping process from problem statement to multiprocessor implementation as a series of transformations from one stage of algorithm specification to another.

The user first defines a *problem statement* and investigates various *algorithm approaches*. The resulting algorithm is called the *virtual algorithm* and its best known implementation is called the *ideal algorithm*. (For example, for an image-processing problem, the user may determine that the discrete fourier transform is an effective approach for one part of the solution and incorporate it into the virtual algorithm. The ideal algorithm would employ the fast fourier transform for this procedure). This process, which is part of the *algorithm development life cycle*, can be iterated multiple times.

The virtual and ideal algorithms possess an identifiable structure which may be characterized in terms of a basic set of features such as *type of parallelism, data granularity, synchronization*, etc. These features may be compared to the characteristics of a given multiprocessor architecture or a stored library of multiprocessor architectures. In the first case, matching the features of the algorithm with the characteristics of the given multiprocessor will help determine how easily the algorithm may be supported on the target machine. This determines how much overhead there will be, and thus whether efficiency is preserved. (If not, another iteration of the algorithm development life cycle should be considered). In the second case, a multiprocessor whose characteristics most closely match algorithm features may be determined to minimize overhead and optimize performance. Figure 1 shows the general relationship between algorithm and architecture characteristics defined by Jamieson [18]. A "1" entry in the table indicates the most

**Architecture Characteristics**

| Algorithm Characteristics | Number of PEs | Memory Organization | Memory Size | Mode (SIMD/MIMD/Pipe) | Network | Synchronization | Processor Capability | Data Types | Addressing Modes | Data Structures | I/O |
|---|---|---|---|---|---|---|---|---|---|---|---|
| Type of Parallelism | 3 | 2 | 3 | 1 |  | 3 |  |  |  |  |  |
| Data Granularity |  | 1 | 2 | 1 | 3 |  |  |  |  | 2 |  |
| Module Granularity |  | 1 |  | 1 | 1 | 1 |  |  |  |  |  |
| Degree of Parallelism | 1 | 2 |  | 2 |  |  |  |  |  |  |  |
| Memory | 2 | 2 | 1 |  |  |  |  |  |  |  |  |
| Uniformity |  | 2 |  | 1 | 3 |  |  |  |  |  |  |
| Synchronization |  | 2 |  | 1 | 2 | 1 |  |  |  |  |  |
| Static/Dynamic |  | 1 |  | 1 | 2 | 3 |  |  |  |  |  |
| Data Dependencies |  | 2 |  | 3 | 1 |  |  |  |  |  |  |
| Fundamental Ops. |  | 2 |  |  | 2 |  | 1 |  |  |  |  |
| Data Types |  |  |  | 3 |  |  |  | 1 | 2 |  | 1 |
| Data Structures |  | 2 |  | 3 | 2 |  |  |  | 3 | 1 |  |
| I/O |  | 3 | 3 | 2 | 2 |  |  |  |  |  | 1 |

*Figure 1*

Jamieson's table [18] of the general relationship between architecture and algorithm characteristics.

dependence between the algorithm and architecture characteristics, a "3" entry indicates less dependence.

The determination of these characteristics for given virtual and ideal algorithms facilitates the implementation of the ideal (or virtual) algorithm on the selected multiprocessor. This implementation is called the *architecture-dependent algorithm*.

Jamieson's work focuses on describing and relating algorithm features and multiprocessor characteristics. The progress from one level of specification to another is determined by the user and when appropriate, the multiprocessor selection process is done using a stored library of multiprocessors and mappings. This work provides architectural and algorithmic models which incorporate general information about execution behavior. Note that with some modification, the architectural model could legitimately be utilized as a "type architecture" [32]. The characterization of algorithm and architectural features relevant to mapping is a strong contribution of this approach.

**The language-based approach**

The language-based approach presupposes that the algorithm is represented by program text and that the mapping compiles or transforms this text to some intermediate form executable on a multiprocessor. Research of this type is often identified with the work of Kuck et. al on the Parafrase compiler ([20], [21], [26]) or Allen and Kennedy on the PFC vectorizer [2]. We illustrate this approach with the work of Kuck and colleagues.

*Parafrase* [21] is a vectorizing compiler which conducts a sequence of source-to-source passes on Fortran 8X program text. The sequence of passes is divided into 3 subsequences: front-end optimizations, intermediate optimizations and back-end optimizations. The front-end optimizations transform the source code for a general high-speed architecture. Intermediate optimizations transform the source code for a more specific class of multiprocessors (typically one of the architecture classes Multiple Execution stream Scalar (MES), Single Execution stream Array (SEA) or Multiple Execution stream Array (MEA) [20]). The back-end optimizations are machine-dependent and transform the source for a specific target architecture.

The *catalog* of available optimizations expose parallelism by restructuring data dependencies in the source code. Data dependencies are determined by analyzing both local information involving pairs of statements in the program and global information involving information flow through entire subroutines. The data dependencies are taxonomized as *flow dependent, anti-dependent, output dependent* and for some compilers, *control dependent*. This approach to data analysis is more accurate and less conservative than conventional data flow analysis. Kuck and colleagues have found that for typical numeric Fortran programs, array subscripts are often dependent on loop indices and constants which can be analyzed at compile-time. For many of these programs, the data analysis is computationally efficient with near-linear time complexity; hence this approach to data analysis leads to improved source code and program performance.

The Parafrase project has made multiprocessor execution available to many users unable or unwilling to reformulate their sequential Fortran code. In addition, this work has contributed to our understanding of vectorizing compilers and to the detection of parallelism in conventional programming language constructs.

### The linear algebra-based approach

The linear algebra-based approach is often used in the VLSI/systolic array multiprocessor environment. The algorithm is represented by a set of algebraic objects which typically includes a matrix of data dependencies. Algebraic transformations are applied to the algorithm representation to determine a mapping into the multiprocessor representation. Good examples of this research are the work of Kung [22], Capello and Steiglitz [10], Varman and Ramakrishnan [37], and Fortes and Moldovan ([15], [24]). We illustrate this approach with Fortes and Moldovan's work.

Fortes and Moldovan map (sequential) programs with bounded loops into fixed-size systolic arrays. The algorithm is represented as a 5-tuple of indices, computations, input and output variables, and dependencies. The indices represent loop iteration and statement numbers. The dependencies are typed as *self-dependencies, input dependencies* and *internal dependencies*. (The data dependencies in this work, as in the Parafrase project, represent more information than in conventional dataflow analysis). The target multiprocessor is envisioned as a VLSI systolic array characterized by a set of interconnection primitives. Fortes and Moldovan's strategy is to use a sequence of linear transformations to map the algorithm dependency matrix into the matrix of interconnection primitives representing the target multiprocessor.

The first set of transformations are used to derive an *execution ordering* on the columns (dependencies) in the dependency matrix. The execution ordering identifies the first component of each dependency as its time component and must be chosen to ensure that all operands of a computation are available before its execution.

The execution ordering places no limit on the number of computations which may be executed simultaneously. Hence, the next transformations are selected to minimize execution time and map indices (computations) into the processors of the target array. The indices may be partitioned using a set of *time hyperplanes* and *partitioning hyperplanes* which divide the index space into *bands*. All indices on the same time hyperplane may be processed in parallel; indices on the same partitioning hyperplane must be processed sequentially. (See Figure 2).

Once the time and partitioning hyperplanes are determined, a policy for scheduling the bands and the computations within them is determined. All indices inside one band must be processed before another band is considered. (Communications between computations in distinct bands are performed via external FIFO queue registers which temporarily store variables). Note that the restricted algorithm domain and this set of transformations enable the user to precisely determine the execution time of the given algorithm on the array. This mapping strategy been implemented as the ADVIS system at USC.

Fortes and Moldovan's work provides a rigorous mathematical framework in which to describe and evaluate algorithms, architectures and mappings. Sufficient conditions for the execution of large sets of independent computations and the correctness of their execution orderings may be guaranteed. Note that this is a "depth-first" mapping strategy, i.e. by restricting the algorithm domain to a subdomain of sequential looping programs, Fortes and Moldovan, like the Parafrase group, are able to focus on performing the mapping particularly well for this domain.

**The graph-based approach**

The graph-based approach represents both the algorithm and architecture models in terms of graph structures and the mapping in terms of graph partitions and/or embeddings. This approach has been used successfully by Bokhari [8], Fishburn and Finkel [14], Stone [35], DeGroot [12], Browne [9], Cuny and Bailey [11] and a number of other researchers. Our work on the Prep-P project ([3], [4]) utilizes a graph-based approach.

We illustrate the graph-based approach with some recent work by Sarkar and Hennessy in mapping dataflow graphs to shared and non-shared memory MIMD architectures ([29], [28]). Algorithms are represented in this work by dataflow computation graphs in the IF1 format [31]. An IF1 program is composed of a hierarchy of acyclic dataflow graphs whose nodes are either simple (atomic) or compound. Sarkar and Hennessy map programs in this format into an MIMD multiprocessor model using a sequence of four

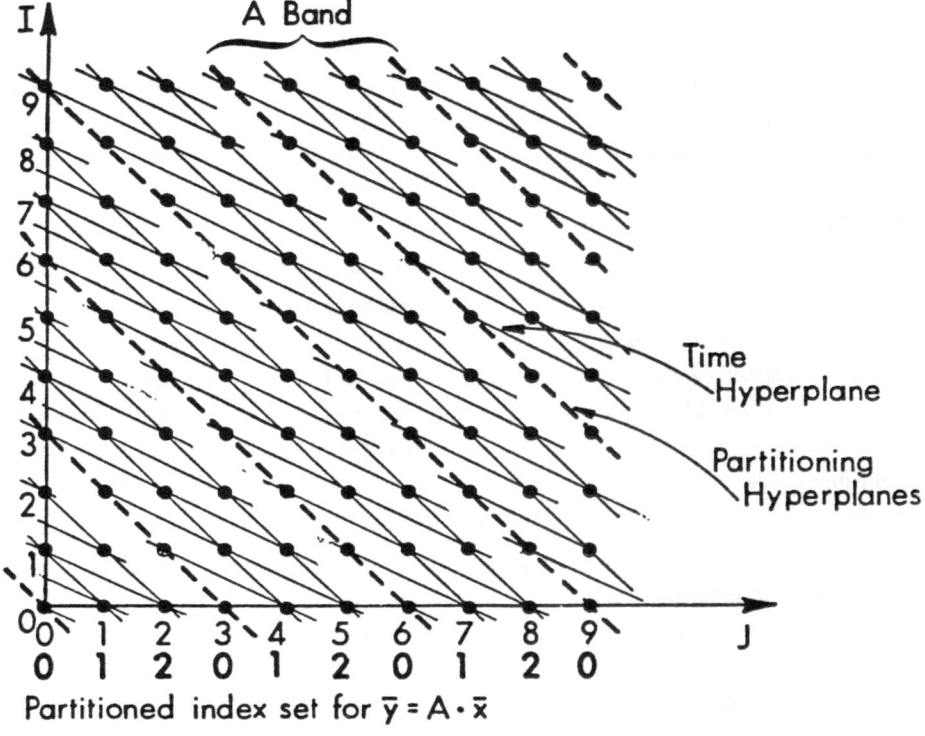

*Figure 2*

Fortes and Moldovan's [24] depiction of the mapping of a two-dimensional index space representing a matrix-vector computation. The indices lie on *time hyperplanes* representing the computations which may be executed on a 3 processor linear array simultaneously. *Partitioning hyperplanes* relate computations in time. All computations in a single band must be scheduled before computations in another band. In this example, the scheduling policy is to process time hyperplanes from the bottom up and bands from left to right.

transformations: *cost assignment, graph expansion, internalization* and *processor assignment*. We will briefly describe each of these transformations.

The first transformation assigns execution time costs to the nodes and communication time costs to the edges of the program graph *(cost assignment)*. These costs may be assessed heuristically, by programmer directive, or by using profile data. The cost assignment transforms the program graph into a weighted program graph from which parallel tasks may be identified.

The next transformation recursively expands the compound computation nodes in the weighted program graph into subgraphs of simpler tasks *(graph expansion)*. This process completes when the nodes in the expanded graph are atomic or are smaller than a given bound on the execution time. The execution time bound is determined as a function of the number of processors in the target multiprocessor, the total program cost and a *granularity threshold factor*.

The expanded task graph then undergoes a transformation called *internalization*. During internalization, the task graph is partitioned into blocks such that all tasks in a single block will be executed by a single processor. (A convexity constraint is utilized to ensure that the introduction of blocks does not create cycles in the task block graph). The difficulty in constructing this partition is to balance the time complexity of parallel task execution with the overhead of interprocessor communication. This problem is NP-complete [16] and requires good heuristics in practice. (Note that the algorithm domain is restricted so the problem is decidable). Sarkar and Hennessy have implemented this system [28] and use a greedy algorithm for this portion of the software.

Finally, the blocks in the task block graph are assigned to processors in the target machine *(processor assignment)*. The assignment is made to minimize parallel execution time, another NP-complete problem. In practice, Sarkar and Hennessy use a priority list scheduling algorithm [1] to perform the processor assignment.

Sarkar and Hennessy's work combines the graph-based algorithm representation with a mapping model which includes a substantive amount of program information. The mapping process is clean and efficient and yet employs both computation- and communication- balancing heuristics for algorithms in its program domain. The dependence of the partitioning and embedding process on specific program behavior is a strong contribution of this work.

## The Prep-P mapping approach

It is within the graph-based approach that our work on the mapping problem lies. For us, the program domain is the set of decomposable parallel programs which can be represented by an interconnection topology of communicating sequential processes. The multiprocessor model is a non-shared memory MIMD model. The mapping methodology focuses on distinguishing the problems of size and structure mismatch between the communication topologies of parallel algorithms and the processor interconnection structures of their target machines. The goal is to provide a performance-efficient, automatic mapping from the algorithm representation into the multiprocessor model. The implementation of this method is called the *Prep-P* system.

Prep-P is based upon the mapping methodology described in [6], [7]: A series of transformations are performed on the algorithm communication graph to map it into the interconnection graph of the target multiprocessor. These transformations *contract* the algorithm graph onto an intermediate graph, and *place* and *route* the intermediate graph in the target interconnection architecture. This provides a graph embedding which first eliminates the size and then the structure mismatch between the parallel algorithm and the target machine. In practice, the mapped graph is then *multiplexed* on the target architecture so that after execution, the termination and output behavior of the mapped algorithm would be the same as if the original parallel algorithm were run on a multiprocessor which exactly matched its size and structure. Figure 3 gives an example of the mapping of a 15 node summing algorithm into a 9 processor multiprocessor lattice using these transformations.

The Prep-P software follows the methodology outlined above. Programs written in a special graph description language are *contracted* (when appropriate), *placed, routed* and *multiplexed*. The output is assembly language code executable on the Poker simulator [34] of the CHiP machine [33]. (The CHiP machine is a non-shared memory reconfigurable parallel architecture which can simulate a variety of MIMD interconnection networks such as the hypercube, shuffle-exchange, omega, butterfly, mesh, torus, binary tree, etc. [36]).

The *contraction* portion of the software partitions the input communication topology into a set of at most P blocks (where P is the number of available processors). The goal is to maximize parallel execution and to minimize the number of inter-block communications. This problem is NP-complete and thus requires good heuristics. Similar to Sarkar and Hennessy's approach, we weight the program graph with communication and computation costs. We derive these costs from a straightforward preprocessing of the code to determine rough estimates of communication load and computation time. The weighted program graph is then used in conjunction with a modified local neighborhood search algorithm [1] to find a good partitioning.

The *placement* and *routing* transformations take the contracted graph and embed it on a CHiP lattice. The placement algorithm is based upon the Kernighan and Lin algorithm [19] and the routing algorithm is a simple breadth-first search [1]. Both algorithms are currently being modified to utilize the weights derived for the contraction algorithm and external I/O information. We expect the resulting Prep-P "front-end" to produce performance-efficient mappings which are sensitive to program behavior.

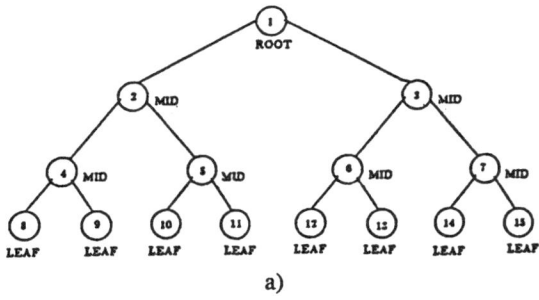

a)

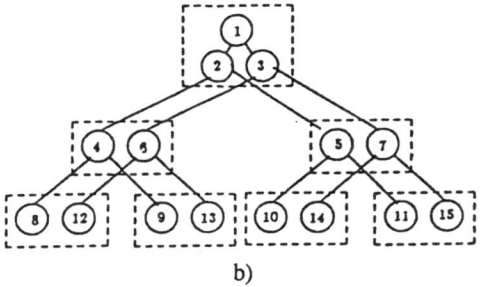

b)

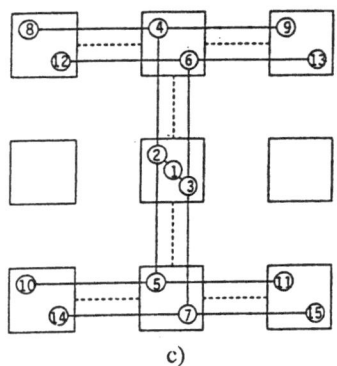

c)

*Figure 3*

Mapping of a 15 node summing algorithm into a 9 processor lattice. The algorithm inputs a data stream from off-chip to each leaf which is then passed to the leaf's parent. At each iteration, the internal nodes sum the data coming in from their left and right children and pass the results to their parents. The root computes the global sum and finally passes the data off-chip. The algorithm communication structure is a broadcast tree shown in Figure 3a). An intermediate contracted graph is shown in Figure 3b). The placed and routed version of the contracted graph on the 9 processor lattice is shown in Figure 3c).

The *multiplexing* portion of the Prep-P software (the "back-end") concatenates process codes from the original parallel algorithm mapped to the same processor and includes routines which manage external, inter- and intra-process I/O and context-switch from one process to the other. The Prep-P preprocessor is currently under design and development by our group here at UCSD ([3], [17], [27], [5]), and will be available (from us) for distribution in 1988.

Note that although Sarkar and Hennessy's method and the Prep-P methodolgy are both graph-based approaches, they differ on several key points. First, the algorithm representation in each case encodes quite different information. IF1 graphs give explicit representation of the dataflow and are composed of processes which may be expanded during execution. In Prep-P input graphs, the dataflow is implicit in the communication topology and processes are treated as atomic. IF1 graphs generally describe the domain of programs which can be represented within single-assignment languages like Sisal [23]; Prep-P input graphs generally describe the domain of parallel algorithms which may be decomposed into a set of communicating sequential processes.

The contrast between algorithm representations dictates different applicable transformations. IF1 graphs are amenable to graph expansion so that the internalization transformation operates on an expanded version of the input graph. Further, each process is expected to run to completion before another can be started in the final task block graph. Prep-P graphs are not amenable to graph expansion so that contraction algorithms operate on a weighted version of the input graph. In addition, although processes are considered atomic during the mapping process, the multiprocessor is expected to context-switch between processes mapped to the same processor during execution. The mapping transformations for both approaches must take such differences into account.

### Section 3: Summary

As stated in the introduction, the work described here is only the "tip of the iceberg" of mapping research. We expect research within each of these approaches to continue, and to contribute to the understanding of algorithm and multiprocessor models and to the development of high-level parallel programming environments. It is interesting to note, however, several similarities between the different approaches.

Each approach incorporates some mechanism by which the mapping process is sensitive to program behavior. For example, the researchers within the linear algebra- and language-based approaches described here utilize a more detailed form of dependency analysis than conventional dataflow analysis. This analysis enables more parallelism to be extracted from the input code, often with a performance-efficient implementation. The graph-based researchers utilize estimates of communication and computation costs of a given algorithm to derive program-sensitive mappings. This approach produces different (and better) mappings for distinct algorithms with the same communication topologies. The characteristics-based approach involves taxonomizing algorithm features to obtain the best match to multiprocessor features. Program behavior manifests itself in the determination of such characteristics as type of parallelism, uniformity, module granularity, etc. Each of these approaches utilize specific information about program behavior in deriving the mappings.

In addition, many researchers are implementing their mapping methodologies to determine performance and provide mapping tools. Our experience on the Prep-P project has shown the development process to be extremely valuable; the interplay between methodology and implementation enhances both activities, guiding the design of mapping models and improving software tools. Software tools also bring this research from the literature to the general multiprocessor user community.

A problem common to each of the approaches is to provide a rigorous but general definition of what is meant by an "optimal" mapping. Researchers agree that optimal mappings must minimize parallel execution time but other characteristics depend heavily on the target multiprocessor. For example, within the graph-based approach, the time and overhead differential between external, inter- and intra-processor communication substantively affect the choice of an optimal mapping. In the linear algebra-based approach, minimal parallel execution time may be exhibited by several mappings, each of which may involve different patterns of data usage and require more or less external I/O. Some researchers have studied the optimality problem [25] but more work must be done in each approach.

Note also that there is some commonality of algorithm domain within and between approaches. Researchers within the graph-based approach typically focus on some subdomain of parallel algorithms while researchers within the language-based and linear algebra-based approaches typically focus on sequential programs. General-purpose mapping methodologies should accommodate both program domains. In addition, most of the research mentioned here focuses on static algorithms of some type. The mapping of dynamic algorithm domains into multiprocessors remains for the most part an important open problem.

Currently, researchers within the parallel computation community continue to study and define solutions to the mapping problem. Their research not only contributes to the design of expressive algorithm and multiprocessor models but also to the development of high-level parallel programming environments. Moreover, the software systems implementing these mapping strategies render multiprocessing more accessible to practitioners as well as researchers. The importance of the mapping problem, and the usefulness of its solution to both researchers and users enhance the task of the researcher in this challenging and intriguing area.

## Acknowledgements

I would like to thank John Rice and the Institute for Mathematics and its Applications for inviting me to the Workshop where a preliminary version of this paper was given. Thanks also to M. E. Bock, Shahid Bokhari, John Feo, Jeanne Ferrante, Charles Holland, Leah Jamieson, Dave Mizell, Dan Moldovan, H. J. Siegel, Steve Skedziewelski, Larry Snyder and Bernd Stramm for enlightening and enjoyable discussions on the mapping problem.

# REFERENCES

[1]     AHO, A., HOPCROFT, J. and J. ULLMAN, *Data Structures and Algorithms*, Addison Wesley, 1983.

[2]     ALLEN, J. and K. KENNEDY, *PFC: A Program to Convert Fortran to Parallel Form*, Proceedings of the IBM Conference on Parallel Computers and Scientific Computations, March, 1982.

[3]     BERMAN, F., *Experience with an Automatic Solution to the Mapping Problem*, The Characteristics of Parallel Algorithms, edited by L. Jamieson, D. Gannon and R. Douglass, MIT Press, Cambridge, Mass., 1987.

[4]     BERMAN, F., GOODRICH, M., KOELBEL, C., ROBISON, W. and K. SHOWELL, *Prep-P: A Mapping Preprocessor for CHiP Architectures*, Proceedings of the 1985 International Conference on Parallel Processing, Pheasant Run, Illinois.

[5]     BERMAN, F. et al., Prep-P Internal Documentation, unpublished project documentation.

[6]     BERMAN and L. SNYDER, *Mapping Parallel Algorithms into Parallel Architectures*, extended abstract, Proceedings of the 1984 International Conference on Parallel Processing, Bellaire, Michigan.

[7]     BERMAN, F. and L. SNYDER, *Mapping Parallel Algorithms into Parallel Architectures*, Journal of Parallel and Distributed Computing, to appear.

[8]     BOKHARI, S., *On the Mapping Problem*, IEEE Transactions on Computers, Vol. C-30, March, 1981.

[9]     BROWNE, J., *Formulation and Programming of Parallel Computations: A Unified Approach*, Proceedings of the 1985 International Conference on Parallel Processing, Pheasant Run, Illinois.

[10]    CAPELLO, P. and K. STEIGLITZ, *Unifying VLSI Array Design with Linear Transformations of Space-Time*, Advances in Computing Research, Vol. 2, JAI Press, 1984.

[11]    CUNY, J. and D. BAILEY, *Graph Grammar Based Specification of Interconnection Structures*, Department of Computer and Information Science Technical Report A-86-23, University of Massachussetts at Amherst, July, 1986.

[12]    DEGROOT, D., *Partitioning Job Structures for SW-Banyan Networks*, Proceedings of the 1983 International Conference on Parallel Processing, Bellaire, Michigan.

[13]    DELP, E., SIEGEL, H.J., WHINSTON, A. and L. JAMIESON, *An Intelligent Operating System for Executing Image Understanding Tasks on a Reconfigurable Parallel Architecture*, Proceedings of the IEEE Computer Society Workshop on Computer Architecture for Pattern Analysis and Image Database Management, Miami, Fl., 1985.

[14]    FISHBURN, J. and R. FINKEL, *Quotient Networks*, IEEE Transactions on Computers, Vol. C-31, April, 1982.

[15]    FORTES, J. and D. MOLDOVAN, *Parallelism Detection and Transformation Techniques Useful for VLSI Algorithms*, Journal of Parallel and Distributed Computing, Vol. 2, August, 1985.

[16]    GAREY, M. and D. JOHNSON, *Computers and Intractability, A Guide to NP-Completeness*, W.H. Freeman and Company, 1979.

[17]    HADEN, P. and F. BERMAN, *A Comparative Study of Mapping Algorithms for an Automated Parallel Programming Environment*, Department of Computer Science Technical Report CS-088, University of California, San Diego.

[18]    JAMIESON, L., *Characterizing Parallel Algorithms,* The Characteristics of Parallel Algorithms edited by L. Jamieson, D. Gannon and R. Douglass, MIT Press, Cambridge, Mass., 1987.

[19]    KERNIGHAN, B. and S. LIN, *An Efficient Heuristic Procedure for Partitioning Graphs,* Bell System Technical Journal, February, 1970.

[20]    KUCK, D. et. al, *The Effects of Program Restructuring, Algorithm Change and Architecture Choice on Program Performance,* Proceedings of the 1984 International Conference on Parallel Processing, Bellaire, Michigan.

[21]    KUCK, D., KUHN, R., LEASURE, B. and M. WOLFE, *The Structure of an Advanced Retargetable Vectorizer,* Tutorial on Supercomputers, Designs and Applications, edited by K. Hwang, IEEE Press, New York, 1984.

[22]    KUNG, S.Y. *On Supercomputing with Systolic/Wavefront Array Processors,* Proceedings of the IEEE, Vol. 72, July, 1984.

[23]    MCGRAW, J., SKEDZIELEWSKI, S. et. al, *SISAL: Streams and Iteration in a Single-Assignment Language,* M-146, Lawrence Livermore National Laboratory, March, 1985.

[24]    MOLDOVAN, D. and J. FORTES, *Partitioning Algorithms for Fixed Size VLSI Architectures,* Department of Electrical Engineering-Systems Technical Report PPP 83-5, University of Southern California, 1983.

[25]    NELSON, P. and L. SNYDER, *Programming Solutions to the Algorithm Contraction Problem,* Proceedings of the 1986 International Conference on Parallel Processing, Pheasant Run, Illinois.

[26]    PADUA, D. and M. WOLFE, *Advanced Compiler Optimizations for Supercomputers,* Communications of the ACM, Volume 29, Number 12, December, 1986.

[27]    ROSE, D. and F. BERMAN, *Mapping with External I/O: A Case Study,* Proceedings of the 1987 International Conference on Parallel Processing, Pheasant Run, Illinois.

[28]    SARKAR, V., *Partitioning and Scheduling Parallel Programs for Execution on Multiprocessors,* Ph.D. Dissertation, Computer Systems Laboratory Technical Report No. CSL-TR-87-328, Stanford University, April 1987.

[29]    SARKAR, V. and J. HENNESSY, *Compile-time Partitioning and Scheduling of Parallel Programs,* Proceedings of the 1986 SIGPLAN Symposium on Compiler Construction.

[30]    SEGALL, Z. and L. SNYDER, editors, *Proceedings of the NSF-CMU Workshop on Performance-Efficient Parallel Programming,* Seven Springs, Champion, Pa., Department of Computer Science Technical Report, Carnegie-Mellon University, 1987.

[31]    SKEDZIELEWSKI, S. and J. GLAUERT, *IF1 - An Intermediate Form for Applicative Languages,* Version 1.0, M-170, Lawrence Livermore National Laboratory, July, 1985.

[32]    SNYDER, L., *Type Architectures, Shared Memory, and the Corollary of Modest Potential,* Annual Review of Computer Science, 1:289-317, 1986.

[33]    SNYDER, L., *Introduction to the Configurable, Highly Parallel Computer,* Computer, January, 1982.

[34]    SNYDER, L., *Introduction to the Poker Parallel Programming Environment,* Proceedings of the 1983 International Conference on Parallel Processing, Bellaire, Michigan.

[35]    STONE, H., *Parallel Processing with the Perfect Shuffle,* IEEE Transactions on Computers, C-20:2, 1971.

[36]   ULLMAN, J., *Computational Aspects of VLSI*, Computer Science Press, 1984.

[37]   VARMAN, P. and I. RAMAKRISHNAN, *On Matrix Multiplication Using Array Processors*, Lecture Notes in Computer Science: Proceedings of the 12th International Conference on Automata, Languages and Programming, Springer-Verlag, Volume 194, July, 1985.

# APPLICATIONS OF GRÖBNER BASES IN NON-LINEAR COMPUTATIONAL GEOMETRY

## BRUNO BUCHBERGER[1]

### Abstract

*Gröbner bases* are certain finite sets of multivariate polynomials. Many
problems in polynomial ideal theory (algebraic geometry, non-linear computational geometry) can be solved by easy algorithms after transforming the
polynomial sets involved in the specification of the problems into Gröbner basis form. In this paper we give some examples of applying the Gröbner bases
method to problems in non-linear computational geometry (inverse kinematics
in robot programming, collision detection for superellipsoids, implicitization
of parametric representations of curves and surfaces, inversion problem for
parametric representations, automated geometrical theorem proving, primary
decomposition of implicitly defined geometrical objects). The paper starts with
a brief summary of the Gröbner bases method.

# 1 Introduction

Traditionally, computational geometry deals with geometrical and *combinatorial*
problems on *linear objects* and simple non-linear objects, see for example (Preparata,
Shamos 1985). These methods are not appropriate for recent advanced problems
arising in geometrical modeling, computer-aided design, and robot programming,
which are more *algebraic* in nature and involve *non-linear geometrical objects*. Real
and complex algebraic geometry is the natural framework for most of these non-
linear problems. Unfortunately, in the past decades, algebraic geometry was very
little concerned with the algorithmic solution of problems. Rather, *non-constructive*
proofs of certain geometrical phenomena and mere existence proofs for certain geometrical objects was, and still is, the main emphasis.

The method of Gröbner bases is an *algorithmic* method that can be used to
attack a wide range of problems in commutative algebra (polynomial ideal theory)
and (complex) algebraic geometry. It is based on the concept of Gröbner bases
and on an algorithm for constructing Gröbner bases introduced in (Buchberger
1965, 1970). In recent years the method has been refined and analyzed and more

---

[1]RISC-LINZ (Research Institute for Symbolic Computation) Johannes Kepler University, A4040
Linz, Austria. This research is supported by a grant from VOEST-ALPINE, Linz, (Dipl. Ing. H.
Exner), and a grant from SIEMENS, München, (Dr. H. Schwärtzel)

applications have been studied. (Buchberger 1985) is a tutorial and survey on the Gröbner bases method.

The present paper starts with a brief summary of the *basic concepts and results of Gröbner bases theory* ( Section 2). If the reader accepts these basic concepts and results as black boxes, the main part of the paper is self-contained. The internal details of the black boxes together with extensive references to the literature are given in the tutorial (Buchberger 1985).

The main part of the paper explains various applications of the Gröbner bases method for problems in non-linear computational geometry as motivated by advanced *applications* in computer-aided design, geometrical modeling and robot programming. The sequence for the presentation of these applications is quite random. Each of them relies on one or several of the basic properties of Gröbner bases summarized in Section 2.

# 2    Summary of the Gröbner Bases Method

The reader who is interested only in the applications may skip this section and come back in case he needs a specific notation, concept or theorem.

## 2.1    General Notation

| | |
|---|---|
| **N** | set of natural numbers including zero |
| **Q** | set of rational numbers |
| **R** | set of real numbers |
| **C** | set of complex numbers |
| $K$ | typed variable for arbitrary fields |
| $\overline{K}$ | algebraic closure of $K$ |
| $i, j, k, l, m, n$ | typed variables for natural numbers |
| $K[x_1, \ldots, x_n]$ | ring of $n$-variate polynomials over the coefficient field $K$ |
| $K(x_1, \ldots, x_n)$ | field of $n$-variate rational rational expressions over the coefficient field $K$ |
| $a, b, c, d$ | typed variables for elements in coefficient fields |
| $f, g, h, p, q$ | typed variables for polynomials |
| $s, t, u$ | typed variables for power products, i. e. polynomials of the form $x_1^{i_1} \ldots x_n^{i_n}$ |
| $C(f, u)$ | the coefficient at power product $u$ in polynomial $f$ |
| $F, G$ | typed variables for finite sets of polynomials |
| $H$ | typed variable for finite sequences of polynomials |
| Ideal($F$) | the ideal generated by F, i. e. the set $\{\sum_i h_i.f_i \mid h_i \in K[x_1, \ldots, x_n], f_i \in F\}$ |
| Radical($F$) | the radical of the ideal generated by F, i. e. $\{f \mid f$ vanishes on all common zeros of $F\}$ or, equivalently, $\{f \mid f^k \in$ Ideal($F$) for some $k\}$ |
| $f \equiv_F g$ | $f$ is congruent to $g$ modulo Ideal($F$) |
| $K[x_1, \ldots, x_n]/$Ideal($F$) | the residue class ring modulo Ideal(F) |
| $[f]_F$ | the residue class of $f$ modulo Ideal($F$) |

In the definition of Ideal($F$), it is sometimes necessary to explicitly indicate the polynomial ring from which the $h_i$ are taken. If the polynomial ring is not clear from the context, we will use an index:

$\text{Ideal}_{K[x_1,\ldots,x_n]}(F)$

In the definition of Radical($F$), by a common zero of the polynomials in F we mean a common zero in the algebraic closure of the coefficient field.

## 2.2 Polynomial Reduction

The basic notion of Gröbner bases theory is *polynomial reduction*. The notion of polynomial reduction depends on a linear ordering on the set of power products that can be extended to a partial ordering on the set of polynomials. The set of *"admissible orderings"* that can be used for this purpose can be characterized by two easy axioms. The *lexical ordering* and the *total degree ordering* are the two admissible orderings used most often in examples. These two orderings are completely specified by fixing a linear ordering on the set of indeterminates $x_1, \ldots, x_n$ in the polynomial ring. Roughly, $f$ reduces to $g$ modulo $F$ iff $g$ results from $f$ by subtracting a suitable multiple $a.u.h$ of a polynomial $h \in F$ such that $g$ is lower in the admissible ordering than $f$. Reduction may be conceived as a generalization of the subtraction step that appears in univariate polynomial division. For all details, see (Buchberger 1985). We use the following notation:

| | |
|---|---|
| $\succ$ | typed variable for admissible orderings |
| $\text{LP}(f)$ | leading power product of $f$ (w. r. t. $\succ$) |
| $\text{LC}(f)$ | leading coefficient of $f$ (w. r. t. $\succ$) |
| $\text{MLP}(F)$ | the set of "multiples of leading powerproducts in $F$", i. e. $\{u \mid (\exists f \in F)(u \text{ is a multiple of } \text{LP}(f)\}$ |
| $f \rightarrow_F g$ | $f$ reduces to $g$ modulo $F$ |
| $\rightarrow^*_F$ | reflexive-transitive closure of $\rightarrow_F$ |
| $\leftrightarrow^*_F$ | reflexive-symmetric-transitive closure of $\rightarrow_F$ |
| $\underline{f}_F$ | $f$ is in normal form modulo $F$, i. e. there does not exist any $g$ such that $f \rightarrow_F g$ |

A binary relation $\rightarrow$ on a set $M$ is called *"noetherian"* iff there does not exist any infinite sequence $x_1 \rightarrow x_2 \rightarrow x_3 \rightarrow \ldots$ of elements $x_i$ in $M$.

### Lemma 2.2.1 (Basic Properties of Reduction)
*(Noetherianity)*
   *For all F:* $\rightarrow_F$ *is noetherian.*
*(Reduction Closure = Congruence)*
   *For all F:* $\equiv_F = \leftrightarrow^*_F$.
*(Normal Form Algorithm)*
   *There exists an algorithm* NF *("Normal Form") such that for all $F, g$:*
   *(NF1)*   $g \rightarrow^*_F \text{NF}(F,g)$,
   *(NF2)*   $\underline{\text{NF}(F,g)}_F$.
*(Cofactor Algorithm)*

*There exists an algorithm* COF *("cofactors") such that for all $F, g$:*
COF$(F, g)$ *is a sequence $H$ of polynomials indexed by $F$ satisfying*
$g = \text{NF}(F, g) + \sum_{f \in F} H_f \cdot f$.

Note that, for fixed $F, f$, there may exist many different $g$ such that $f \to_F^* g$ and $\underline{g}_F$ i. e. , in general, "normal forms for polynomials $f$ are not unique modulo $F$". A normal form algorithm NF, by successive reduction steps, singles out one of these $g$ for each $F$ and $f$.

COF proceeds by "collecting" the multiples $a.u.h$ of polynomials $h \in F$ that are subtracted in the reduction steps when applying the normal form algorithm NF to $g$. Actually, COF can be required to satisfy additional properties, for examples, certain restrictions on the leading power products of the polynomials $H_f \cdot f$.

## 2.3   Gröbner Bases and the Main Theorem

### Definition 2.3.1 (Buchberger 1965, 1970)
*$F$ is a Gröbner basis (w. r. t. $\succ$) iff*
 *"normal forms modulo $F$ are unique", i.e.*
 *for all $f, g_1, g_2$:*
  *if $f \to_F^* g_1, f \to_F^* g_2, \underline{g_1}_F, \underline{g_2}_F$, then $g_1 = g_2$.*

Note that $\to_F$ depends on the underlying admissible ordering $\succ$ on the power products. Therefore, also the definition of Gröbner basis depends on the underlying $\succ$. Whenever $\succ$ is clear from the context, we will not explicitly mention $\succ$. Gröbner bases whose polynomials are monic (i. e. have leading coefficient 1) and are in normalform modulo the remaining polynomials in the basis are called *"reduced Gröbner bases"*. As we will see, Gröbner bases have a number of useful properties that establish easy algorithms for important problems in polynomial ideal theory. Therefore the main question is how Gröbner bases can be algorithmically constructed. The algorithm needs the concept of *"S-polynomials"*. The S-polynomial of two polynomials $f$ and $g$ is the difference of certain multiples $u.f$ and $v.g$. For details see (Buchberger 1985). We use the notation

SP$(f, g)$                    the S-polynomial of $f$ and $g$.

### Theorem 2.3.1 (Main Theorem, Buchberger 1965, 1970)
*(Algorithmic Characterization of Gröbner Bases)*
 *$F$ is a Gröbner basis iff*
  *for all $f, g \in F$ : NF$(F, \text{SP}(f, g)) = 0$.*
*(Algorithmic Construction of Gröbner Bases)*
 *There exists an algorithm GB such that for all $F$*
  *(GB1)   Ideal$(F) = $ Ideal$(\text{GB}(F))$*
  *(GB2)   GB$(F)$ is a (reduced) Gröbner basis.*

The proof of the (Algorithmic Characterization) is completely combinatorial and quite involved. The whole power of the Gröbner bases method is contained in this proof. The algorithm GB is based on the (Algorithmic Characterization), i. e. it involves successive computation of normal forms of S-polynomials. This algorithm is structurally simple. However, it is complex in terms of time and space consumed. In some sense, this is necessarily so because the problems that can be solved by the Gröbner bases method are intrinsically complex as has been shown by various authors. Still, the algorithm allows to tackle interesting and non-trivial practical problems for which no feasible solutions were known by other methods. Also, various theoretical and practical improvements of the algorithm have enhanced the scope of applicability.

## 2.4   The Gröbner Bases Algorithm in Software Systems

The Gröbner bases algorithm GB is available in almost all major computer algebra systems, notably in the SAC-2, SCRATCHPAD II, REDUCE, MAPLE, MACSYMA and muMATH systems. The introduction of (Buchberger, Collins, Loos 1982) contains the addresses of institutions from which these systems can be obtained. In these systems at least the algorithms SP, NF, (COF,) and GB are accessible to the user. In most systems, also a number of other auxiliary routines and variants of these basic algorithms are available and the user can experiment with different coefficient domains, admissible orderings and strategies for tuning the algorithms.

The implementations vary drastically in their efficiency mostly because of the varying amount of theory that has been taken into account. Also, computation time and space depends drastically on the admissible orderings used, on permutations of variables, on treating indeterminates as ring or field variables, on strategies for selecting pairs in the consideration of S-polynomials and on many other factors. Thus if one seriously considers solving problems of the type described in this paper one should try different systems and various orderings, strategies etc.

The rest of the paper is written with the goal in mind that the reader should be able to apply the methods as soon as he has access to an implementation of the basic algorithms NF, COF, SP, and GB viewed as "black boxes".

## 2.5   Properties of Gröbner Bases

In the following theorem we summarize the most important properties of Gröbner bases on which the algorithmic solution of many fundamental problems in polynomial ideal theory (algebraic geometry, non-linear computational geometry) can be based. Actually, not all of these properties are used in the later sections of the paper. However, since the results on Gröbner bases are quite scattered in the literature, the summary may help the reader who perhaps wants to try the Gröbner bases method on new problems. Many of the properties listed in the theorem were already proven in (Buchberger 1965, 1970). Actually the problems that can be solved with the (Residue Class Ring) properties were the starting point for Gröbner bases theory in (Buchberger 1965). The property (Elimination Ideals) is due to (Trinks 1978). The property (Inverse Mappings) is a recent contribution by (Van den Essen 1986) that solves a decision problem that has been open since 1939. (Algebraic Relations) and (Syzygies) seem to have been known already to (Spear 1977). However, it is hard to trace were the proofs appeared for the first time. More references are given in (Buchberger 1985). Most of the proofs of the properties below are immediate consequences of the definition of Gröbner bases, the property (Reduction Closure = Congruence), and some well known algebraic lemmas in polynomial ideal theory. The proofs of the properties (Syzygies) and (Inverse Mappings) are more involved. The existence of the algorithm GB based on the above Main Theorem is the crux for the algorithmic character of the properties.

In the following, let $K[x_1, \ldots, x_n]$ be arbitrary but fixed. $F$ and $G$ are used as typed variables for finite subsets of $K[x_1, \ldots, x_n]$. If not otherwise stated, $\succ$ is arbitrary. When we say "$y$ is a new indeterminate" we mean that $y$ is different from $x_1, \ldots, x_n$. By "$F$ is solvable" we mean that there exists an $n$-tuple $(a_1, \ldots, a_n)$ of elements $a_i$ in the algebraic closure $\overline{K}$ such that $f(a_1, \ldots, a_n) = 0$ for all $f \in F$. Similarly, the expression "$F$ has finitely many solutions" and similar expressions always refer to solutions over the algebraic closure of $K$.

## Theorem 2.5.1 (General Properties of Gröbner Bases)

*(Ideal Equality, Uniqueness of Reduced Gröbner Bases)*

   *For all $F$, $G$:* $\text{Ideal}(F) = \text{Ideal}(G)$ *iff* $\text{GB}(F) = \text{GB}(G)$.

*(Idempotency of GB)*

   *For all reduced Gröbner bases $G$:* $\text{GB}(G) = G$.

*(Ideal Membership)*

   *For all $F$, $f$:* $f \in \text{Ideal}(F)$ *iff* $\text{NF}(\text{GB}(F), f) = 0$.

*(Canonical Simplification)*

   *For all $F$, $f$, $g$:* $f \equiv_F g$ *iff* $\text{NF}(\text{GB}(F), f) = \text{NF}(\text{GB}(F), g)$.

*(Radical Membership)*

   *For all $F$, $f$:*
   $f \in \text{Radical}(F)$ *iff* $1 \in \text{GB}(F \cup \{y.f - 1\})$, *(where $y$ is a new indeterminate)*.

*(Computation in Residue Class Rings)*

   *For all $F$:*

   *The residue class ring $K[x_1, \ldots, x_n]/\text{Ideal}(F)$ is isomorphic to the algebraic structure whose carrier set is $\{f \mid \underline{f}_F\}$ and whose addition and multiplication operations, $\oplus$ and $\otimes$, are defined as follows:*

   $$f \oplus g := \text{NF}(\text{GB}(F), f + g),$$
   $$f \otimes g := \text{NF}(\text{GB}(F), f.g).$$

*(Note that the carrier set is a decidable set and $\oplus$ and $\otimes$ are computable!).*

*(Residue Class Ring, Vector Space Basis)*

*For all F:*

*The set $\{[u]_F \mid u \notin \text{MLP}(\text{GB}(F))\}$ is a linearly independent basis for $K[x_1, \ldots, x_n]/\text{Ideal}(F)$ considered as a vector space over K.*

*(Residue Class Ring, Structure Constants)*

*For all F, u, v:*
  *if $u, v \notin \text{MLP}(\text{GB}(F))$,*
  *then $[u]_F.[v]_F = \sum_{w \notin \text{MLP}(\text{GB}(F))} a_w.[w]_F$,*
    *where, for all $w, a_w := \text{C}(\text{NF}(\text{GB}(F), u.v), w)$.*

*(The $a_w \in K$, appearing in these representations of products of the basis elements as linear combinations of the basis elements are the "structure constants" of $K[x_1, \ldots, x_n]/\text{Ideal}(F)$ considered as an associative algebra.)*

*(Leading Power Products)*

*For all F: $\text{MLP}(\text{Ideal}(F)) = \text{MLP}(\text{GB}(F))$.*

*(Principal Ideal)*

*For all F:*
  *Ideal(F) is principal (i. e. has a one-element ideal basis)*
  *iff GB(F) has exactly one element.*

*(Trivial Ideal)*

*For all F: $\text{Ideal}(F) = K[x_1, \ldots, x_n]$ iff $\text{GB}(F) = \{1\}$.*

*(Solvability of Polynomial Equations)*

*For all F: F is solvable iff $1 \notin \text{GB}(F)$.*

*(Finite Solvability of Polynomial Equations)*

*For all F:*
  *F has only finitely many solutions iff*
  *for all $1 \le i \le n$ there exists an $f \in \text{GB}(F)$ such that*
    *LP(f) is a power of $x_i$.*

*(Number of Solutions of Polynomial Equations)*

*For all F with finitely many solutions:*
  *the number of solutions of F (with multiplicities and solutions at infinity) =*
  *= cardinality of $\{u \mid u \notin \mathrm{MLP}(\mathrm{GB}(F))\}$.*

*(Minimal Polynomial)*

  *For all F and all finite sets U of power products:*
  *There exists an $f \in \mathrm{Ideal}(F)$ in which only power products from U occur
  iff $\{\mathrm{NF}(\mathrm{GB}(F, u)) \mid u \in U\}$ is linearly dependent over K.*

  *(By applying this property successively to the powers $1, x_i, x_i^2, x_i^3, \ldots$ one
  can algorithmically find, for example, the univariate polynomial in $x_i$
  of minimal degree in $\mathrm{Ideal}(F)$ if it exists. On this algorithm a general method for solving arbitrary system of polynomial equations can be
  based, see (Buchberger 1970), which works for arbitrary $\succ$ whereas the
  elimination method mentioned below works only for lexical orderings.)*

*(Syzygies)*

  *Let F be a (reduced) Gröebner basis and define for all $f, g \in F$:*

  $P^{(f,g)} := \mathrm{COF}(F, \mathrm{SP}(f, g)),$
  *u and v such that $\mathrm{SP}(f, g) = u.f - v.g$,*
  $S^{(f,g)}$ *is a sequence of polynomials indexed by F,*
  $S_f^{(f,g)} := u - P_f^{(f,g)},$
  $S_g^{(f,g)} := -v - P_g^{(f,g)},$
  $S_h^{(f,g)} := -P_h^{(f,g)},$ *for all $h \in F - \{f, g\}$.*

*Then,*

  $\{S^{(f,g)} \mid f, g \in F\}$ *is a set of generators for the $K[x_1, \ldots, x_n]$-module of
  all sequences H of polynomials (indexed by F) that are solutions ("syzy-
  gies") of the linear diophantine equation*

  $\sum_{h \in F} H_h.h = 0.$

  *(This solution method for linear diophantine equations over $K[x_1, \ldots, x_n]$
  whose coefficients form a Gröbner basis F can be easily extended to the
  case of arbitrary F and to systems of linear diophantine equations, see
  (Buchberger 1985), (Winkler 1986)).*

**Theorem 2.5.2 (Properties of Gröbner Bases for Particular Orderings)**

*(Hilbert Function)*

*Let $\succ$ be a total degree ordering.*
*Then, for all F:*

*The value $H(d, F)$ of the Hilbert function for d and F, i. e. the number of modulo $\text{Ideal}(F)$ linearly independent polynomials in $K[x_1, \ldots, x_n]$ of degree $\leq d$, is equal to*

$$\binom{d+n}{n} - \text{cardinality of } \{u \text{ of degree} \leq d \mid u \notin \text{MLP}(\text{GB}(F))\}.$$

*(Elimination Ideals, Solution of Polynomial Equations)*

*Let $\succ$ be the lexical ordering defined by $x_1 \prec x_2 \prec \ldots \prec x_1$.*
*Then, for all F, $1 \leq i \leq n$:*

*The set $\text{GB}(F) \cap K[x_1, \ldots, x_i])$ is a (reduced) Gröbner basis for the "i-th elimination ideal" generated by F, i. e. for $\text{Ideal}_{K[x_1,\ldots,x_n]}(F) \cap K[x_1, \ldots, x_i]$.*

*(This property leads immediately to a general solution method, by "successive substitution", for arbitrary systems of polynomial equations with finitely many solutions, which is formally described in (Buchberger 1985). We will demonstrate this method in the examples in the application section of this paper.)*

*(Continuation of Partial Solutions)*

*Let $\succ$ be a lexical ordering.*
*For all F:*

*If $F := \{f_1, \ldots, f_k\}$ is a Gröbner basis with respect to $\succ$, $f_1 \prec \cdots \prec f_k$, and $f_1, \ldots, f_l (1 \leq l \leq k)$ are exactly those polynomials in F that depend only on the indeterminates $x_1, \ldots, x_i$, then every common solution $(a_1, \ldots, a_i)$ of $\{f_1, \ldots, f_l\}$ can be continued to a common solution $(a_1, \ldots, a_n)$ of F. (For a correct statement of this property some terminology about solutions at infinity would be necessary.)*

*(Independent Variables Modulo an Ideal)*

*For all F and $1 < i_1 < \ldots < i_m < n$:*

*The indeterminates $x_{i_1}, \ldots, x_{i_m}$ are independent modulo $\text{Ideal}(F)$ (i. e. there is no polynomial in $\text{Ideal}(F)$ that depends only on $x_{i_1}, \ldots, x_{i_m}$) iff $\text{GB}(F) \cap K[x_{i_1}, \ldots, x_{i_m}] = \{0\}$, where the $\succ$ used must be a lexical ordering satisfying $x_{i_1} \prec \cdots \prec x_{i_m} \prec$ all other indeterminates. (This property yields immediately an algorithm for determining the dimension of a polynomial ideal (algebraic variety).)*

*(Ideal Intersection)*

> Let $\succ$ *be the lexical ordering defined by* $x_1 \prec x_2 \prec \ldots \prec x_n \prec y$,
> $y$ *a new variable.*
> *Then, for all* $F$, $G$:
>
> $\text{GB}(\{y.f \mid f \in F\} \cup \{(y-1).g \mid g \in G\}) \cap K[x_1, \ldots, x_n]$
> *is a (reduced) Gröbner basis for* $\text{Ideal}(F) \cap \text{Ideal}(G)$.
>
> *(This property yields also an algorithm for quotients of finitely generated ideals because the determination of such quotients can be reduced to the determination of intersections.)*

*(Algebraic Relations)*

> *For all* $F$:
>
> > Let $F = \{f_1, \ldots, f_m\}$, *let* $y_1, \ldots, y_m$ *be new indeterminates and let* $\succ$ *be the lexical ordering defined by* $y_1 \prec \ldots \prec y_m \prec x_1 \prec \ldots \prec x_n$.
> > *Then,* $\text{GB}(\{y_1 - f_1, \ldots, y_m - f_m\}) \cap K[y_1, \ldots, y_m]$ *is a (reduced) Gröbner basis for the "ideal of algebraic relations" over* $F$, *i. e. for the set* $\{g \in K[y_1, \ldots, y_m] \mid g(f_1, \ldots, f_m) = 0\}$.

*(Inverse Mapping)*

> *For all* $F$:
>
> > Let $F = \{f_1, \ldots, f_n\}$, *let* $y_1, \ldots, y_n$ *be new indeterminates and let* $\succ$ *be the lexical ordering defined by* $y_1 \prec \ldots \prec y_n \prec x_1 \prec \ldots \prec x_n$. *Then, the mapping from* $\overline{K}^n$ *into* $\overline{K}^n$ *defined by* $F$ *is bijective iff* $\text{GB}(\{y_1 - f_1, \ldots, y_n - f_n\})$ *has the form* $\{x_1 - g_1, \ldots, x_1 - g_n\}$ *for certain* $g_j \in K[y_1, \ldots, y_n]$.

The properties stated in the above theorem can be read as the algorithmic solution of certain problems specified by polynomial sets $F$. Each of these "algorithms" requires that, for solving the problem for an arbitrary $F$, one first transforms $F$ into the corresponding (reduced) Gröbner basis $\text{GB}(F)$ and then performs some algorithmic actions on $\text{GB}(F)$. For example, for the decision problem "$f \equiv_F g$?", (Canonical Simplification) requires that one first transforms $F$ into $\text{GB}(F)$ and then checks, by applying algorithm NF, whether or not the normal forms of $f$ and $g$ are identical modulo $\text{GB}(F)$. Actually, most of the above properties (algorithms) are correct also if, instead of transforming $F$ into a corresponding *reduced* Gröbner basis, one transforms $F$ into an *arbitrary* equivalent Gröbner basis $G$. (We say "$F$ is equivalent to $G$" iff $\text{Ideal}(F) = \text{Ideal}(G)$.) In practice, however, this makes very little difference because the computation of Gröbner bases is not significantly easier if one relaxes the requirement that the Gröbner basis must be reduced.

Alternatively, by (Idempotency of GB), the properties stated in the above theorem can also be read as properties of (reduced) Gröbner bases — and algorithms for solving problems for (reduced) Gröbner bases. For example, introducing the additional assumption that F is a (reduced) Gröbner basis, (Canonical Simplification) reads as follows:

*For all (reduced) Gröbner bases $F$, and polynomials $f, g$:*
   $f \equiv_F g$ *iff* $\mathrm{NF}(F, f) = \mathrm{NF}(F, g)$.

Some of the properties stated in the above theorem are characteristic for Gröbner bases, i. e. if the property holds for a set $F$ then $F$ is a Gröbner basis. For example, (Leading Power Products) is a characteristic property, i. e. if $\mathrm{MLP}(\mathrm{Ideal}(F)) = \mathrm{MLP}(F)$ then F is a Gröbner basis.

Let us carry through one more exercise for reading the above properties as algorithms. For deciding whether

(Question)
   for all $a_1, \ldots, a_n \in \overline{K}$,
      for which $f_1(a_1, \ldots, a_n) = \cdots = f_m(a_1, \ldots, a_n) = 0$,
      also $g(a_1, \ldots, a_n) = 0$,

i. e. for deciding whether $g \in \mathrm{Radical}(\{f_1, \ldots, f_m\})$, because of (Radical Membership), it suffices to perform the following steps:

1. Compute the (reduced) Gröbner basis $G$ for $\{f_1, \ldots, f_m, y.g - 1\})$,
   where $y$ is a new indeterminate.
2. The (Question) has a positive answer iff $1 \in G$.

# 3   Application: Inverse Robot Kinematics

The problem of inverse robot kinematics is the problem of determining, for a given robot, the distances at the prismatic joints and the angles at the revolute joints that will result in a given position and orientation of the end-effector. The mathematical description of this problem leads to a system of multivariate polynomial equations (after representing angles $\alpha$ by their sine and cosine and adding $\sin^2 \alpha + \cos^2 \alpha = 1$ to the set of equations), see (Paul 1981).

Let us consider, for example, the following robot having two revolute joints (two "degrees of freedom").

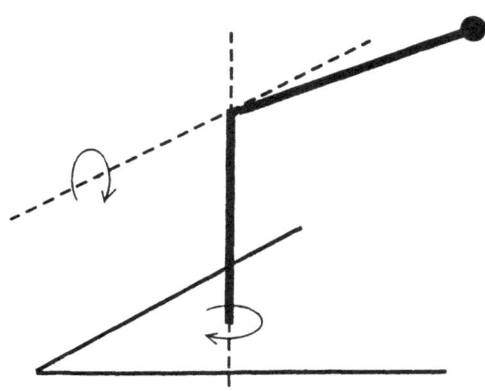

We introduce the following variables:

$l_1, l_2$            lengths of the two robot arms
$px, py, pz$        $x$-, $y-$, and $z$-coordinate of the position of the end-effector
$\phi, \theta, \psi$           Euler angles of the orientation of the end effector
                        (Euler angles are one way of describing orientation)
$\delta_1, \delta_2$           angles describing rotation at the revolute joints

We introduce the sines and cosines of the angles occuring in the above description as separate variables:

$s_1, c_1$            sine and cosine of $\delta_1$
$s_2, c_2$            sine and cosine of $\delta_2$
$sf, cf$              sine and cosine of $\phi$
$st, ct$               sine and cosine of $\theta$
$sp, cp$             sine and cosine of $\psi$

The interrelation of the physical entities described by the above variables is expressed in the following system of equations:

$$c_1 \cdot c_2 - cf \cdot ct \cdot cp + sf \cdot sp = 0,$$
$$s_1 \cdot c_2 - sf \cdot ct \cdot cp - cf \cdot sp = 0,$$
$$s_2 + st \cdot cp = 0,$$
$$-c_1 \cdot s_2 - cf \cdot ct \cdot sp + sf \cdot cp = 0,$$
$$-s_1 \cdot s_2 + sf \cdot ct \cdot sp - cf \cdot cp = 0,$$
$$c_2 - st \cdot sp = 0,$$
$$s_1 - cf \cdot st = 0,$$
$$-c_1 - sf \cdot st = 0,$$
$$ct = 0,$$
$$l_2 \cdot c_1 \cdot c_2 - px = 0,$$
$$l_2 \cdot s_1 \cdot c_2 - py = 0,$$
$$l_2 \cdot s_2 + l_1 - pz = 0,$$
$$c_1^2 + s_1^2 - 1 = 0,$$
$$c_2^2 + s_2^2 - 1 = 0,$$
$$cf^2 + sf^2 - 1 = 0,$$
$$ct^2 + st^2 - 1 = 0,$$
$$cp^2 + sp^2 - 1 = 0.$$

Let us call those variables that describe the geometrical realization of the robot "geometrical variables" (for example, the variables $l_1, l_2$). Let us also call those variables that describe position and orientation of the end-effector shortly "position variables" ($px, \ldots, sf, cf, \ldots$). The other variables ($s_1, c_1, \ldots$) are the "joint variables".

In the case of more complicated robots (with six degrees of freedom), one can specify values for the geometrical variables and the position variables and, with certain restrictions, will always be able to determine appropriate values of the joint variables that yield the given position and orientation of the end-effector. In the above example robot, with only two degrees of freedom however one can only independently choose the value of two position variables, for example $px$ and $pz$. The value of all the other variables, notably of the other position variables $py, sf, cf, \ldots$, and the joint variables will then be determined by the above system of algebraic equations.

The problem can be considered in three different versions of increasing generality.

(Real Time Version)

- The value of the geometrical variables are numerically given.

- The value of those position variables that can be independently chosen (e. g. $px, pz$) are numerically given.

- The solution of the problem consists in determining appropriate numerical values for the (remaining position variables and) the joint variables.

(Off-Line Version, Concrete Robot)

- The value of the geometrical variables are numerically given.

- The value of those position variables that can be independently chosen are left open as *parameters*.

- By a "solution of the problem", in this version, one means symbolic expressions involving the position parameters that describe, in "closed form", the dependence of the (remaining position variables and) the joint variables from the position parameters. Of course, a "symbolic closed form solution" of this kind will not always be possible. It is possible for certain classes of robots, see (Paul 1981), and it is possible in a modified sense also in the general case by using Gröbner bases.

(Off-Line Version, Robot Class)

- The value of the geometrical variables are left open as *parameters*.

- The value of those position variables that can be independently chosen are left open as *parameters*.

- By a "solution of the problem", in this version, one means symbolic expressions involving the geometrical and the position parameters that describe, in "closed form", the dependence of the (remaining position variables and) the joint variables on the geometrical *and* position parameters.. A "symbolic closed form solution" in this general sense is even more difficult. Again, it is possible for certain classes of robots and, as we shall see, it is possible in a modified sense also in the general case by using Gröbner bases.

A symbolic solution of the inverse kinematics problem in the (Off-Line Version), can be contrasted to a numerical approach:

(Symbolic Approach)

- Derivation of the symbolic expressions for the solution of the problem in the (Off-Line Version).

- Numerical specification of the parameters.

- Numerical evaluation of the symbolic expressions using the numerical values of the parameters.

(Numerical Approach)

- Numerical specification of the parameters.

- Solution of the problem in the (Real-Time Version) by numerical iteration methods.

It is clear that a symbolic solution of the problem in the (Off-Line Version) can have practical advantages over the purely numerical approach (as long as the resulting symbolic expressions describing the solutions are not too complicated) because the numerical evaluation of the symbolic solution expressions in real-time situations may be faster than a direct iterative numerical solution of the (Real Time Version) of the problem. Also, of course, the symbolic solution may give "insight" into the problem that can not be gained by a numerical solution.

For the above example, we show the solution of the problem in the (Off-Line Version, Roboter Class) by using Gröbner bases. In this version, the geometrical variables $l_1, l_2$ and the position variables $px, pz$ are considered as symbolic parameters.

The solution method uses property (Elimination Ideals) of Gröbner bases. This property, read as an algorithm, tells us that we first have to compute the Gröbner bases of the set $F$ of input polynomials. Since $l_1, l_2, px, pz$ are to be treated as symbolic parameters, we work over the field $Q(l_1, l_2, px, pz)$ as coefficient field. This is perfectly possible, because the Gröbner bases method works over arbitrary fields (whose arithmetic is algorithmic). Furthermore, we must specify an ordering on the remaining variables, for example $c_1 \prec c_2 \prec s_1 \prec s_2 \prec py \prec cf \prec ct \prec cp \prec sf \prec st \prec sp$. These variables are treated as ring variables, i.e. the Gröbner basis will be computed considering the input polynomials as polynomials in the ring $Q(l_1, l_2, px, pz)[c_1, \ldots, sp]$. The resulting Gröbner basis has the following form:

$$c_1^2 + \frac{px^2}{pz^2 - 2 \cdot l_1 \cdot pz - l_2^2 + l_1^2} = 0,$$

$$c_2 + \frac{pz^2 - 2 \cdot l_1 \cdot pz - l_2^2 + l_1^2}{l_2} \cdot px \cdot c_1 = 0,$$

$$s_1^2 - \frac{pz^2 - 2 \cdot l_1 \cdot pz + px^2 - l_2^2 + l_1^2}{pz^2 - 2 \cdot l_1 \cdot pz - l_2^2 + l_1^2} = 0,$$

$$s_2 - \frac{pz - l_1}{l_2} = 0,$$

$$py + \frac{pz^2 - 2 \cdot l_1 \cdot pz - l_2^2 + l_1^2}{px} \cdot c_1 \cdot s_1 = 0,$$

$$cf^2 - \frac{pz^2 - 2 \cdot l_1 \cdot pz + px^2 - l_2^2 + l_1^2}{pz^2 - 2 \cdot l_1 \cdot pz - l_2^2 + l_1^2} = 0,$$

$$ct = 0,$$

$$cp + \frac{pz^3 - 3 \cdot l_1 \cdot pz^2 - l_2^2 \cdot pz + 3 \cdot l_1^2 \cdot pz + l_1 \cdot l_2^2 - l_1^3}{l_2 \cdot pz^2 - 2 \cdot l_1 \cdot l_2 \cdot pz + l_2 \cdot px^2 - l_2^3 + l_1^2 \cdot l_2} \cdot s_1 \cdot cf = 0,$$

$$sf + \frac{pz^2 - 2 \cdot l_1 \cdot pz - l_2^2 + l_1^2}{pz^2 - 2 \cdot l_1 \cdot pz + px^2 - l_2^2 + l_1^2} \cdot c_1 \cdot s_1 \cdot cf = 0,$$

$$st + \frac{pz^2 - 2 \cdot l_1 \cdot pz - l_2^2 + l_1^2}{pz^2 - 2 \cdot l_1 \cdot pz + px^2 - l_2^2 + l_1^2} \cdot s_1 \cdot cf = 0,$$

$$sp + \frac{pz^4 - 4 \cdot l_1 \cdot pz^3 - 2 \cdot l_2^2 \cdot pz^2 + 6 \cdot l_1^2 \cdot pz^2 + 4 \cdot l_1 \cdot l_2^2 \cdot pz - 4 \cdot l_1^3 \cdot pz + l_2^4 - 2 \cdot l_1^2 \cdot l_2^2 + l_1^4}{l_2 \cdot px \cdot pz^2 - 2 \cdot l_1 \cdot l_2 \cdot px \cdot pz + l_2 \cdot px^3 - l_2^3 \cdot px + l_1^2 \cdot l_2 \cdot px} \cdot c_1 \cdot s_1 \cdot cf = 0.$$

The above Gröbner basis has a remarkable structure:

- The geometrical parameters $l_1$ and $l_2$ and the position parameters $px$ and $pz$ are still available as symmbolic parameters in the polynomials of the Gröbner basis. Thus, the system is still "general". The Gröbner basis is in "closed form".

- In accordance with property (Elimination Ideals), the system is "triangular-ized". In this example, this means that the first polynomial of the basis depends only on $c_1$, the second on $c_1, c2$, the third on $c_1, c_2, s_1, \ldots$. After substitution of numerical values for the parameters $l_1, l_2, px, pz$, we can therefore numerically determine the possible values for $c_1$ from the first equation then, for each of the values of $c_1$, determine the value of $c_2$ from the second equation then, for each of the values of $c_1, c_2$, determine the value of $s_1$ from the third equation etc.

- Actually, the degrees of the polynomials in this basis are quite low. This is in general not true for the first polynomial in Gröbner bases. The first polynomial, which, in case the solution set is finite, is always univariate, tends to have quite a high degree in general. The degrees of the other polynomials, however, tend to be very low (most times even linear) also in the general case because the polynomial sets describing realistic physical or geometrical situations often define prime ideals, for which linearity in the second, third ... variable can be proven theoretically. This phenomenon needs closer study, however. For numerical practice, low degrees in the second, third ... variable implies that numerical errors from the determination of the value of the first value will not drastically accumulate. In the case where the second, third ... equation is linear, the Gröbner basis has the form $\{p_1(x_1), x_2 - p_2(x_1), \ldots, x_n - p_n(x_1)\}$. In this case, the errors introduced by the numerical solution of $p_1$ will not accumulate at all.

- The above method of numerical backward substitution based on the Gröbner basis, by property (Elimination Ideals), is guaranteed to yield *all* (real and complex) solutions of the system.

- Again by (Elimination Ideals), no *"extraneous"* solutions of the system are produced. (Other algebraic methods, for example the resultant method, may produce extraneous solutions.)

The above Gröbner basis was produced in 62 sec on an IBM 4341 using an implementation of the Gröbner basis method by R. Gebauer and H. Kredel in the SAC-2 computer algebra system. The computation time is increasing drastically when more complicated robot types are investigated. We are far from being able to treat the most general robot of six degrees of freedom. However, so far, only very little research effort has been dedicated to this possible application of Gröbner bases. Using the special structure of the problem it may well be that more theoretical results can be derived that allow to drastically speed up the general algorithm in this particular application.

# 4   Application: Intersection of Superellipsoids

Superellipsoids (Barr 1981) are surfaces in 3D space that have a compact implicit representation as the set of points $(x, y, z)$ such that

$$((\frac{x}{a})^{2/\epsilon_3} + (\frac{y}{b})^{2/\epsilon_2})^{\epsilon_3/\epsilon_1} + (\frac{z}{c})^{2/\epsilon_1} - 1 = 0$$

Superellipsoids are topologically equivalent to spheres. They can be considered as ellipsoids with axes $a, b, c$ whose curvature in the $x-, y-, z-$ directions is distorted by the influence of the exponents $\epsilon_1, \epsilon_2, \epsilon_3$. (The above equation is the implicit equation for the case where the superellipsoid is in standard position with its midpoint at the origin.) The exponents $\epsilon_1, \epsilon_2, \epsilon_3$ open an enormous flexibility for adjusting the shape of superellipsoids in order to approximate real objects. Some basic problems in geometric modeling, for example, the problem of deciding whether a point is inside or outside an object can be easily solved for superellipsoids. Recently, superellipsoids have been proposed for approximating parts of robots and obstacles in order to test for collision. The collision detection problem of robots is thereby reduced to an intersection test for superellipsoids.

Unfortunately, for general superellipsoids, no good intersection tests are known. In this section we report on first attempts to apply Gröbner bases for this question. We restrict our attention to the case of a sphere (with midpoint $(A, B, C)$ and radius $R$) and a superellipsoid (in standard position) whose exponents satisfy $\epsilon_1 = \epsilon_2 = \epsilon_3 < 2$ (a convex superellipsoid). In this case, the two objects intersect iff the minimal distance between the midpoint of the sphere and the superellipsoid is less or equal to the radius of the sphere. Using Lagrange factors, this approach leads to the following system of equations for the coordinates $(x, y, z)$ of the point on the superellipsoid having minimal distance to $(A, B, C)$:

(Equations for Minimal Distance)

$(\frac{x}{a})^{2/\epsilon} + (\frac{y}{b})^{2/\epsilon} + (\frac{z}{c})^{2/\epsilon} - 1 = 0$

$(x - A) + \lambda.\frac{1}{\epsilon.a}.(\frac{x}{a})^{(2/\epsilon-1)} = 0$

$(y - B) + \lambda.\frac{1}{\epsilon.b}.(\frac{y}{b})^{(2/\epsilon-1)} = 0$

$(z - C) + \lambda.\frac{1}{\epsilon.c}.(\frac{z}{c})^{(2/\epsilon-1)} = 0$

If $\epsilon$ is of the form $1/k$ (which is sufficiently general for practical purposes), this (System for Minimal Distance) is an algebraic system. We consider $a, b, c, A, B, C$ as parameters, i. e. we work over $K(a, b, c, A, B, C)[x, y, z, \lambda]$. For computing the Gröbner bases, we use the lexical ordering defined by $x \prec y \prec z \prec \lambda$. For $\epsilon = 1$ (which is, actually, the ellipsoid case) we get the Gröbner basis

(Gröbner Basis for Minimal Distance)

$x^6 - p(x) = 0$

$y - q(x) = 0$

$z - r(x) = 0$

$\lambda - s(x) = 0.$

Here, $p(x), q(x), r(x), s(x)$ are univariate polynomials in $x$ of degree 5 with coefficients that are rational expressions in the parameters $a, b, c, A, B, C$. The equation for $\lambda$ is not interesting for the problem at hand and may be dropped. The printout of these rational expressions consumes approximately 2 pages. (Some simplification by extracting common subexpressions would be possible.) Again, the Gröbner basis has all the advantageous features described in the inverse kinematics application. Note in particular that, in this Gröbner basis, the second, third and fourth equations are linear in the variables $y, z, \lambda$, respectively. Therefore the Gröbner basis presents an explicit symbolic solution to the problem as soon as the solution value for $x$ is numerically determined from the first equation, which is univariate in $x$.

If we change $\epsilon$ to $1/2$, the resulting Gröbner basis will again have the structure displayed in (Gröbner Basis for Minimal Distance). The only difference is that the degree of the univariate polynomials $p(x), q(x), r(x), s(x)$ will be 11. We conjecture that the structure of the system will stay unchanged for arbitrary $\epsilon$ of the form $1/k$.

The problem with this approach is, again, computation time. While the Gröbner basis computation for $\epsilon = 1$ needs 15 minutes (on an IBM 4341 in the SAC-2 implementation of the Gröbner bases method), the computation already needs 19 hours for $\epsilon = 1/2$. At the moment, this excludes practical applicability of the method. However, one should take into account that the source of complexity seems to be the extraneous extremal solutions that enter through the Lagrange factor method. Actually, the first equation in the Gröbner basis describes the $x$-coordinate of all relative extremal points on the surface and not only the $x$-coordinates of the minimal point. This raises the degree of the first polynomial and, hence, also of the other polynomials. More systematic study is necessary. Furthermore, it seems to be possible to guess and subsequently prove the general structure of the polynomials $p(x), q(x), r(x), s(x)$ from the Gröbner bases computations for two or three different $\epsilon$ values. This could make the Gröbner basis computation superfluous in the future. As with other symbolic computation methods, Gröbner bases computations can be applied on very different levels including the level of producing and supporting mathematical conjectures.

# 5 Application: Implicitization of Parametric Objects

As has been pointed out repeatedly, the automatic transition between implicit and parametric representation of curves and surfaces is of fundamental importance in geometric modeling, see for example (Sederberg, Anderson 1984). The reason for this is that the implicit and the parametric representation are appropriate for different classes of problems. For example, for generating points along curves or surfaces, the parametric representation is most convenient whereas, for deciding whether a given point lies on a specific curve or surface, the implicit representation is most natural. It is also well known that implicitization of parametric surfaces is of importance for deriving a representation of the intersection curve of two surfaces. This problem has a satisfactory solution in case one of the surfaces is expressed parametrically and the other implicitly. In this case, the parameter representation $x(s,t), y(s,t), z(s,t)$ for the first surface can be substituted into the implicit equation $f(x,y,z)$ of the other surface. This results in the implicit representation $f(x(s,t), y(s,t), z(s,t))$ of the intersection curve in parameter space.

Actually, for some time, the problem of implicitization has been deemed unsolvable in the CAD literature. (Sederberg, Anderson 1984), however, presented a solution of the implicitization problem using resultants. The solution is spelled out for surfaces in 3D and curves in 2D. In the general case of $(n-1)$-dimensional hypersurfaces, I guess, the method could yield implicit equations that introduce non-trivial extraneous solutions, see also the remarks in (Arnon, Sederberg 1984). In (Arnon, Sederberg 1984) it is shown how Gröbner bases can be used for the general implicitization problem of $(n-1)$-dimensional hypersurfaces. The authors sketch a correctness proof for the method that relies on (Algebraic Relations). In this section, we review their method and generalize it to the most general case of hypersurfaces of arbitrary dimension in $n$-dimensional space. Still, much research will be needed to assess the efficiencies of the methods and to determine their range of practical applicability. Also some theoretical details are not yet completely covered in the literature.

(General Implicitization Problem)

Given: $p_1, \ldots, p_m \in K[x_1, \ldots, x_n]$.

Find: $f_1, \ldots, f_k \in K[y_1, \ldots, y_m]$,

such that for all $a_1, \ldots, a_m$:

$f_1(a_1, \ldots, a_m) = \cdots = f_k(a_1, \ldots, a_m) = 0$ iff

$a_1 = p_1(b_1, \ldots, b_n), \ldots, a_m = p_m(b_1, \ldots, b_n)$ for some $b_1, \ldots, b_n$.

The problem requires to construct $k$ polynomials implicitly defining hypersurfaces whose intersection is the hypersurface described by the parameter representation.

(Implicitization Algorithm)

$\{f_1, \ldots, f_k\} := \text{GB}(\{y_1 - p_1, \ldots, y_m - p_m\}) \cap K[y_1, \ldots, y_m]$,

where GB has to be computed using the lexical ordering determined by

$y_1 \prec \cdots \prec y_m \prec x_1 \prec \cdots \prec x_n$.

Correctness Proof: Let $g_1 \prec \ldots \prec g_l$ be the polynomials in

$$\text{GB}(\{y_1 - p_1, \ldots, y_m - p_m\}) - K[y_1, \ldots, y_m].$$

$\{y_1 - p_1, \ldots, y_m - p_m\}$ and the Gröbner basis $\{f_1, \ldots, f_k, g_1, \ldots, g_l\}$ have the same common zeros. If

$$f_1(a_1, \ldots, a_m) = \cdots = f_k(a_1, \ldots, a_m) = 0$$

then, by (Continuation of Partial Solutions), there exist $(b_1, \ldots, b_n)$ such that

$$g_1(a_1, \ldots, a_m, b_1, \ldots, b_n) = \cdots = g_l(a_1, \ldots, a_m, b_1, \ldots, b_n) = 0.$$

Hence, also

$$a_1 - p_1(b_1, \ldots, b_n) = 0, \ldots, a_1 - p_1(b_1, \ldots, b_n) = 0.$$

The converse is clear.

**Example:** Let us consider the 3D surface defined by the following parametric representation

(Parametric Representation)

$$x = r.t$$
$$y = r.t^2$$
$$z = r^2$$

Roughly, this surface has the shape of a ship hull whose keel is the $y$-axis and whose bug is the $z$-axis. Applying algorithm GB to $\{x - r.t, y - r.t^2, z - r^2\}$ with respect to the ordering $z \prec y \prec x \prec t \prec r$ yields the following Gröbner basis:

(Gröbner Basis)

$$x^4 - y^2.z$$
$$t.x - y$$
$$t.y.z - x^3$$
$$t^2.z - x^2$$
$$r.y - x^2$$
$$r.x - t.z$$
$$r.t - x$$
$$r^2 - z$$

The polynomial depending only on $x, y, z$ is an implicit equation for the surface defined by (Parameter Representation).

By close inspection one will detect that, actually, the implicit equation occurring in the above (Gröbner Basis) does not strictly meet the specification of the (Implicitization Problem). The $y$-axis is a solution to the implicit equation whereas it does not appear in the surface defined by the (Parameter Representation). This is not a deficiency of the Gröbner basis method but has to do with the particular (Parameter Representation) which, in some sense, is not "general enough" or, stated differently, in the (Continuation of Partial Solutions) property, solutions at infinity have to be taken into account. This question deserves some further detailed study. (Sturmfels 1987) has already sketched some analysis of this phenomenon. He proposes the following parameter presentation, which includes the y-axis and whose implicit equation is again $x^4 - y^2.z$.

(Parametric Representation)

$$x = u.v$$
$$y = v^2$$
$$z = u^4$$

This example was computed in 4 sec on an IBM AT in the author's research implementation of the Gröbner basis method in the muMATH system. Other examples with more complicated coefficients and similar degree characteristics had computing times in the range of several seconds. I guess that the examples occurring in practice should be well tractible by the method.

**Example:** The method can also be used for rational parametric representations. We consider the example of a circle in the plane.

(Rational Parametric Representation)

$x = \frac{1-s^2}{1+s^2}$

$y = \frac{2.s}{1+s^2}$

In the case of rational parametric representations, we first clear denominators. In the example, the input to GB should therefore be $\{x + x.s^2 - 1 + s^2, y + y.s^2 - 2.s\}$. The result is, of course, $x^2 + y^2 - 1$.

# 6 Application: Inversion of Parametric Representations

The inversion problem for parametric representations is defined as follows:

(Inversion Problem for Parametric Representations)
    Given:    $p_1, \ldots, p_m \in K[x_1, \ldots, x_n]$ and
             a point $(a_1, \ldots, a_m)$ on the hypersurface
             parametrically defined by $p_1, \ldots, p_m$.
    Find:     $\{(b_1, \ldots, b_n) \mid a_1 = p_1(b_1, \ldots, b_n), \ldots, a_m = p_m(b_1, \ldots, b_n)\}$.

This problem is closely connected with the (Implicitization Problem). In fact, the (Inversion Problem) is just a special case of the general problem of solving systems of polynomial equations, which is completely solved by the Gröbner basis method based on the (Elimination Ideals) property or based on the (Minimal Polynomial) property. For solving the (Inversion Problem), the general Gröbner bases solution method can be applied to the system $\{y_1 - p_1(x_1, \ldots, x_n), \ldots, y_m - p_m(x_1, \ldots, x_n)\}$, i. e. we have the following algorithm.

(Inversion Algorithm for Parametric Representations)
    $G := \mathrm{GB}(\{y_1 - p_1(x_1, \ldots, x_n), \ldots, y_m - p_m(x_1, \ldots, x_n)\},$
        where GB has to be computed using the lexical ordering determined by
        $y_1 \prec \cdots \prec y_m \prec x_1 \prec \cdots \prec x_n.$
    $\{f_1, \ldots, f_k\} := G \cap K[y_1, \ldots, y_m].$
    (If, for some $1 \leq i \leq k$, $f_i(a_1, \ldots, a_m) \neq 0$, then "Input Error".)
    Substitute $a_i$ for $y_i$ in G and solve the system G, which is "triangularized".

In fact, the steps necessary in this algorithm include the steps of the (Implicitization Algorithm). Therefore, when we apply the Gröbner bases method to the (Implicitization Problem), we automatically get also a solution for the (Inversion Problem) and vice versa.

**Example:** We use again the example of Section 5.

(Parametric Representation)
    $x = r.t$
    $y = r.t^2$
    $z = r^2$

Suppose we want to determine the parameter values defining the point $(2, 2, 4)$ on the surface. Application of GB yields

(Gröbner Basis)

$$x^4 - y^2.z$$
$$t.x - y$$
$$t.y.z - x^3$$
$$t^2.z - x^2$$
$$r.y - x^2$$
$$r.x - t.z$$
$$rt - x$$
$$r^2 - z$$

The first polynomial is the implicit equation, which can be used to check whether $(2, 2, 4)$ is, in fact, on the surface: $2^4 - 2^2.4 = 0$. Substituting $(2, 2, 4)$ in the second, third, and fourth polynomial of the Gröbner basis (and making all polynomials monic) yields the system

(Gröbner Basis After First Substitution)

$$t - 1$$
$$t - 1$$
$$t^2 - 1$$

This system of univariate polynomials, by the property (Continuation of Partial Solutions) must always have a common zero that can be determined by forming the greatest common divisor, $g := t - 1$, of the three polynomials and solving for $t$. This leads to $t = 1$.

Substituting $(2, 2, 4, 1)$ in the fifth, ...,eights polynomial of the Gröbner basis (and making all polynomials monic) yields the system

$$r - 2$$
$$r - 2$$
$$r - 2$$
$$r^2 - 4$$

Again, this system of univariate polynomials, by the property (Continuation of Partial Solutions) must have a common zero that can be determined by forming the greatest common divisor, $h := r - 2$, of the four polynomials and solving for $r$. This leads to $r = 2$.

Actually, it has been shown recently in (Kalkbrener 1987) and, independently, in (Gianni 1987) that the computation of greatest common divisors is not necessary in the above procedure. Rather, as can be verified in the above example, for each of the univariate systems the first non-zero polynomial will always be the greatest common divisor of the system. This is a drastic simplification of the general procedure for solving arbitrary systems of polynomial equations by the Gröbner bases method.

# 7   Application: Detection of Singularities

In tracing implicitly given planar curves, numerical methods work well except when tracing curves through singular points, see (Hofmann 1987). (Hofmann 1987a) has pointed out that Göbner bases yield an immediate approach to detect all singular points of implicitly given planar curves. The singular points of a planar curve given by $f(x, y) = 0$ are exactly the points $(a, b)$ that are common zeros of $f$, $f_x$, and $f_y$. Hence, the problem of determining the set $S$ of singular points of a planar curve $f$ can be treated by the following algorithm.

(Algorithm for Detection of Singularities)

$G := \text{GB}(\{f, f_x, f_y\})$, where $f_x$, $f_y$ are the partial derivatives of $f$ w. r. t. $x$ and $y$ respectively and GB has to be computed w. r. t. a lexical ordering of $x, y$.

$S :=$ set of common zeros of $G$ determined by the successive substitution method.

**Example:** Let us consider the following planar curve

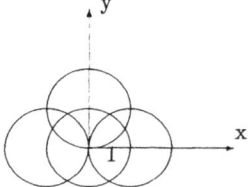

This curve has 9 singular points. We detect them by applying GB to $\{f, f_x, f_y\}$, where

(Four Circle Curve)
$$f := (x^2 + y^2 - 1).((x - 1)^2 + y^2 - 1)((x + 1)^2 + y^2 - 1)(x^2 + (y - 1)^2 - 1).$$

Application of GB, using the lexical ordering determined by $x \succ y$, yields

(Gröbner Basis for Four Circle Curve)
$$y^5.p(y),$$
$$x.y.p(y),$$
$$x^2 - y^4.q(y),$$
$$\text{where} \quad p(y) := y^4 - \tfrac{3}{2}y^3 - \tfrac{1}{4}y^2 - \tfrac{9}{8}y - \tfrac{3}{8},$$
$$q(y) := \tfrac{2558}{27}y^4 - \tfrac{823}{9}y^3 - \tfrac{3895}{54}y^2 + \tfrac{823}{12}y + \tfrac{5}{4}.$$

One sees that, for any solution $y$ of the first polynomial in the Gröbner basis, the second polynomial vanishes identically whereas the third equation yields at most two different values for $x$. Proceeding by the general substitution method for Gröbner bases, we obtain the following singular points:

$(-1, 1), (1, 1),$
$(-1/2, \sqrt{3}/2), (1/2, \sqrt{3}/2),$
$(-\sqrt{3}/2, 1/2), (\sqrt{3}/2, 1/2),$
$(0, 0),$
$(-1/2, -\sqrt{3}/2), (1/2, -\sqrt{3}/2),$

In accordance with the picture, we obtained five different values for $y$ and, altogether, nine singular points. The computation took 78 sec in the author's muMATH Gröbner bases package on an Apollo workstation emulation of an IBM AT.

# 8  Application: Geometrical Theorem Proving

Automated Geometrical Theorem Proving is intriguing in two ways. First, it is a playground for developing and studying new algorithmic techniques for automated mathematics and, second, it becomes more and more important for advanced geometric modeling, which requires to check plausibility and consistency of inaccurate and numerically distorted geometrical objects and to derive and restore their consistent shape, see for example (Kapur 1987). Apart from older approaches to geometrical theorem proving based on heuristics, recently there have been developed three systematic approaches based on three different algorithmic methods in computer algebra, namely Collins' cylindrical algebraic decomposition method (Collins 1975), Wu's method of characteristic sets (Wu 1978) and the Gröbner basis method. (Kutzler 1987) compares the three methods. The use of Gröbner bases for automated geometrical theorem proving has been independently introduced by B. Kutzler and D. Kapur, see for example (Kutzler, Stifter 1986) and (Kapur 1986). In this section we give an outline of the main idea how Gröbner bases can be used for proving geometrical theorems. We start with an example of a geometrical theorem. For simplicity, we present Kapur's approach, Kutzler's approach is slightly different.

**Example:** Apollonios' Circle Theorem.

*The altitude pedal of the hypotenuse of a right-angled triangle and the midpoints of the three sides of the triangle lie on a circle.*

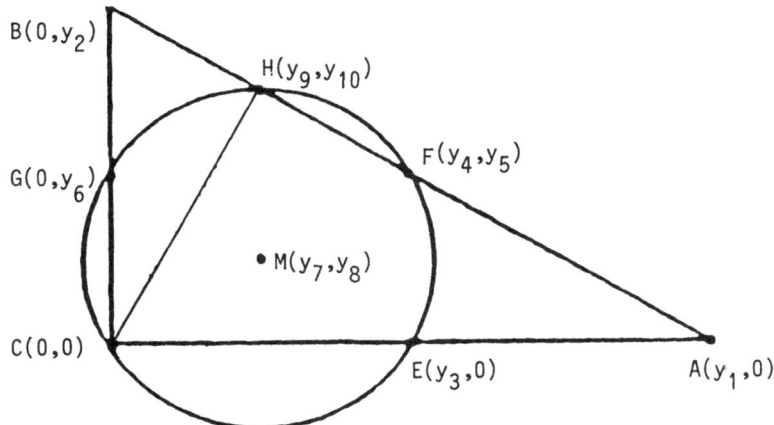

After introducing coordinates, a possible algebraic formulation of this problem is as follows:

(Hypotheses)

$$h_1 := 2y_3 - y_1 = 0 \qquad (E \text{ is midpoint of } CA),$$
$$h_2 := 2y_4 - y_1 = 0 \qquad (F \text{ is midpoint of } AB, \text{ 1st coordinate}),$$
$$h_3 := 2y_5 - y_2 = 0 \qquad (F \text{ is midpoint of } AB, \text{ 2nd coordinate}),$$
$$h_4 := 2y_6 - y_2 \qquad (G \text{ is midpoint of } BC),$$
$$h_5 := (y_7 - y_3)^2 + y_8^2 - (y_7 - y_4)^2 - $$
$$\qquad -(y_8 - y_5)^2 = 0 \qquad (\text{length } EM = \text{length } FM),$$
$$h_6 := (y_7 - y_3)^2 + y_8^2 - (y_8 - y_6)^2 - $$
$$\qquad -y_7^2 = 0 \qquad (\text{length } EM = \text{length } GM),$$
$$h_7 := (y_9 - y_1)y_2 + y_1 y_{10} = 0 \qquad (H \text{ lies on } AB),$$
$$h_8 := -y_1 y_9 + y_2 y_{10} = 0 \qquad (CH \text{ is perpendicular to } AB).$$

(Conclusion)

$$c := (y_7 - y_3)^2 + y_8^2 - (y_7 - y_9)^2 - $$
$$\qquad -(y_8 - y_{10})^2 = 0 \qquad (\text{length } EM = \text{length } HM).$$

To prove the theorem means to show that

for all $a_1, \ldots, a_{10} \in \mathbf{R}$:

if $\quad h_1(a_1, \ldots, a_{10}) = 0, \ldots, h_8(a_1, \ldots, a_{10}) = 0,$

then $c(a_1, \ldots, a_{10}) = 0.$

All expressions $h_i$ and $c$ occurring in this proposition are polynomial expressions. If one replaces $\mathbf{R}$ by $\mathbf{C}$, the proposition, by definition, is just the proposition "$c \in \text{Radical}(\{h_1, \ldots, h_8\})$". However, by (Radical Membership), arbitrary radical membership questions "$c \in \text{Radical}(\{h_1, \ldots, h_m\})$?" can be decided by deciding "$1 \in \text{GB}(\{h_1, \ldots, h_m, z.c - 1\})$?", where $z$ must be a new indeterminate.

This method is totally general and automatic for all geometrical theorems whose hypothesis and conclusions are polynomial equations. In fact, it is also efficient. Hundreds of non-trivial theorems have been proven by this approach, most of them in only several seconds of computing time, see (Kutzler, Stifter 1986), (Kapur 1986) and (Kutzler 1987) for extensive statistics.

Two remarks are appropriate. First, replacing $\mathbf{R}$ by $\mathbf{C}$ slightly distorts the problem. Of course, if a geometrical theorem holds over $\mathbf{C}$ then it also holds over $\mathbf{R}$. The reverse is not true in general. It turns out, however, that the geometrical theorems occuring in the mathematical literature are generally true over $\mathbf{C}$. Still, one must bear in mind that, if a negative answer is produced by this method for a given proposition, this does not necessarily mean that the proposition is false over $\mathbf{R}$. It is false over $\mathbf{C}$, it could be still true over $\mathbf{R}$.

Second, most geometrical theorems are only true for the "general" case. It may well happen that they are false for "degenerate" situations, for examples, when circles have zero radius, angles become zero, lines become parallel etc. Geometric theorem proving based on the Gröbner bases method can handle degenerate situations automatically in a very strong sense.

1. In situations where the degenerate situations can be described in the form $d(x_1, \ldots, x_n) \neq 0$, $d$ a polynomial, one can again use a new indeterminate to transform the question into an ideal (and, hence, Gröbner basis) membership question. Namely,

$$\forall z((h(z) = 0 \wedge s(z) \neq 0) \implies c(z) = 0)$$

is equivalent to

$$\exists z, u, v((h(z) = 0 \wedge u.s(z) = 1 \wedge v.c(z) = 1)$$

is equivalent to

$$1 \in \mathrm{GB}(h, u.s - 1, v.c - 1).$$

Using this wellknown transformation technique one can actually show that the Gröbner basis method yields a decision algorithm for the following general class of formulae:

(quantifiers)(arbitrary boolean combination of polynomial equations)

where either all the quantifiers must be existential or they must be universal, and the formulae must be closed, i. e. no free variables may occur.

2. The Gröbner bases approach to geometrical theorem proving can also be modified in such a way that, in case a proposition does not hold in general, the method automatically produces a set of polynomials describing the degenerate cases in which the proposition may be false. Roughly, this can be done, for example, by analyzing the denominators of the coefficients that are produced when Gröbner bases are computed over rational function coefficient fields. Quite some research has been devoted to this question, see (Kutzler 1986) and (Kapur 1986).

# 9 Application: Primary Decomposition

A polynomial ideal is "decomposable" iff it can be represented as the non-trivial intersection of two other polynomial ideals. Geometrically, this corresponds to a representation of the algebraic manifold (set of zeros) of the ideal as the non-trival union of two algebraic manifolds. It is well known in polynomial ideal theory that every polynomial ideal can be decomposed into finitely many ideals that can not be decomposed further ("irreducible components") and that this decomposition is essentially unique. This is the content of the famous Lasker-Noether decomposition theorem, see for example (Van der Waerden 1953). However, the proof of this theorem is non-constructive, i. e. no general algorithmic method is provided that would find, for a polynomial ideal given by a finite basis $F$, the finite bases for its irreducible components.

In more detail, the primary decomposition of a polynomial ideal (algebraic manifold) $I$ (algebraic manifold) not only gives its irreducible parts (the corresponding "prime ideals") $P_i$ but also information about the "multiplicity" of these irreducible parts. This information is contained in the "primary ideals" $Q_i$ corresponding to the prime ideals. Each prime ideal and its corresponding primary ideal implicitly describe the same irreducible algebraic manifold. However, the prime ideal and a corresponding primary ideal may be different. In this case, the primary ideal tells us "how often" the irreducible manifold defined by the prime ideal occurs in the algebraic manifold defined by the given ideal $I$. Summarizing, the algorithmic version of the primary decomposition problem has the following specification (where we use $Z(F)$ for "set of common zeros of F"):

(Primary Decomposition Problem)

Given: $F$.

Find: $G_i, H_i$ such that
the Ideal($G_i$) are primary,
the Ideal($H_i$) are the prime ideals corresponding to Ideal($G_i$),
Ideal($F$) = $\bigcap_i$ Ideal($G_i$),
(i. e. $Z(F) = \bigcup_i Z(G_i)$), and
some minimality conditions are satisfied.

Note that the problem depends on the underlying coefficient field. For example, $x^2 + 1$ is irreducible over **R** but reducible over **C**.

Recently the problem of algorithmic primary decomposition has been completely solved using Gröbner bases. Still, the algorithm for the most general case is not yet implemented in a software system. Complete implementations may be expected for the very near future. A number of papers, of different generality and level of detail, contributed to the recent progress in this area: (Kandri-Rody 1984), (Lazard 1985), (Gianni, Trager, Zacharias 1985), (Kredel 1987).

An exact formulation of the problem and a detailed description of the algorithms, which are quite involved, is beyond the scope of this paper. It should be clear that automatic decomposition of algebraic manifolds (e. g. intersection curves of 3D objects) should be of utmost importance for geometrical modeling where the global analysis of finitely represented objects, as opposed to a mere local numerical evaluation, is more and more desirable in advanced applications. All the algorithms invented for the solution of the primary decomposition problem heavily rely on the basic properties of Gröbner bases as compiled in Theorem 2.5.1 and Theorem 2.5.2, notably on the properties (Elimination Ideals), (Ideal Membership) and properties derived from these properties as, for example, (Intersection Ideal).

For bringing this important research to the attention of the geometric modeling community we present a simple example showing the kind of information obtainable from a primary decomposition.

**Example:** Primary Decomposition of Cylinder/Sphere Intersection.

Let us consider the intersection of a cylinder with radius $r_1$ whose axis coincides with the $x_3$-axis and a sphere with radius $r_2$ and midpoint at the origin. The intersection curve consists of the common zeros of the following two polynomials:

$$F := \{x_1^2 + x_2^2 - r_1^2, x_1^2 + x_2^2 + x_3^2 - r_2^2\}.$$

Depending on whether $r_1 < r_2$, $r_1 = r_2$, or $r_1 > r_2$, the primary decomposition algorithm, over $\mathbf{R}$, yields the following representation of $\text{Ideal}(F)$ as the intersection of primary ideals:

Case $r_1 < r_2$:
$$\text{Ideal}(F) = \quad \text{Ideal}(x_3 + r, x_2^2 + x_1^2 - r_1^2) \cap \text{Ideal}(x_3 - r, x_2^2 + x_1^2 - r_1^2),$$
$$\text{where } r := \sqrt{r_2^2 - r_1^2}.$$
The two primary components are, in fact, prime.

Case $r_1 = r_2$:
$$\text{Ideal}(F) = \quad \text{Ideal}(x_3^2, x_2^2 + x_1^2 - r_1^2).$$
The ideal is already primary with corresponding prime ideal
$$\text{Ideal}(x_3, x_2^2 + x_1^2 - r_1^2).$$

Case $r_1 > r_2$:
$$\text{Ideal}(F) = \quad \text{Ideal}(x_3^2 - r_2^2 + r_1^2, x_2^2 + x_1^2 - r_1^2).$$
The ideal is already primary and identical to the corresponding prime ideal.

In geometrical terms, the above outcome of the primary decomposition algorithm gives us the following information:

Case $r_1 < r_2$: The manifold decomposes in two irreducible components, namely, two horizontal circles of radius $r_1$ with midpoints $(0, 0, \pm r)$. The multiplicity of these circles is one (the primary ideals are identical to their corresponding prime ideals).

Case $r_1 = r_2$: The manifold does not decompose. It consists of the horizontal circle with radius $r_1$ with midpoint $(0, 0, 0)$. However, this circle has to be "counted twice" because, in the primary ideal, there appears the term $x_3^2$ whereas in the prime ideal, which defines the "shape" (i. e. point set) of the manifold, $x_3$ appears only linearly. This corresponds to the geometrical intuition that the intersection curve results from merging, in the limit, the two horizontal circles of case $r_1 < r_2$.

Case $r_1 > r_2$: The manifold does not decompose (over $\mathbf{R}$!). In fact it has no real points. In contrast to the case $r_1 = r_2$, the manifold has multiplicity one because the primary ideal coincides with the prime ideal.

# 10 Conclusions

The Gröbner bases method provides an algorithmic approach to many problems in polynomial ideal theory. We tried to provide some first evidence that the method could be a valuable tool for the progressing needs of geometrical engineering (geometric modeling, image processing, robotics, CAD etc.).

Further research should concentrate on two areas:

- The theoretical problems (for example, solutions at infinity in paremtric representations) occuring in the application of the method to geometrical problems must be completely studied.

- The computational behavior of the method must be improved by obtaining new mathematical results that could hold in the special situations (e. g. kinematics of certain robot classes) in which the method is applied.

Research on efficiency aspects and on geometrical applications of the Gröbner basis method is only at the beginning.

**Acknowledgement.** I am indebted to C. Hofmann, and B. Sturmfels for personal communications I used in this paper. Thanks also to B. Kutzler, R. Michelic-Birgmayr, and S. Stifter for helping in the preparation of some of the examples.

## REFERENCES

D. S. ARNON, T. W. SEDERBERG, 1984. *Implicit Equation for a Parametric Surface by Gröbner Bases.* In: Proceedings of the 1984 MACSYMA User's Conference (V. E. Golden ed.), General Electric, Schenectady, New York, 431–436.

A. H. BARR, 1981. *Superquadrics and Angle-Preserving Transformations.* IEEE Computer Graphics and Applications, 1/1, 11–23.

B. BUCHBERGER, 1965. *An Algorithm for Finding a Basis for the Residue Class Ring of a Zero-Dimensional Polynomial Ideal (German).* Ph. D. Thesis, Univ. of Innsbruck (Austria), Dept. of Mathematics.

B. BUCHBERGER, 1970. *An Algorithmic Criterion for the Solvability of Algebraic Systems of Equations (German).* Aequationes Mathematicae 4/3, 374–383.

B. BUCHBERGER, G. E. COLLINS, R. LOOS, 1982. "Computer Algebra: Symbolic and Algebraic Computation". Springer-Verlag, Vienna - New York.

B. BUCHBERGER, 1985. *Gröbner Bases: An Algorithmic Method in Polynomial Ideal Theory.* In: Multidimensional Systems Theory (N. K. Bose ed.), D. Reidel Publishing Company, Dordrecht - Boston - Lancaster, 184–232.

G. E. COLLINS, 1975. *Quantifier Elimination for Real Closed Fields by Cylindrical Algebraic Decomposition.* 2nd GI Conference on Automata Theory and Formal Languages, Lecture Notes in Computer Science **33**, 134–183.

P. GIANNI, 1987. *Properties of Gröbner Bases Under Specialization.* Proc. of the EUROCAL '87 Conference, Leipzig, 2–5 June 1987, to appear.

P. GIANNI, B. TRAGER, G. ZACHARIAS, 1985. *Gröbner Bases and Primary Decomposition of Polynomial Ideals.* Submitted to J. of Symbolic Computation. Available as manuscript, IBM T. J. Watson Research Center, Yorktown Heights, New York.

C. HOFMANN, 1987. *Algebraic Curves.* This Volume. Institute for Mathematics and its Applications, U of Minneapolis.

C. HOFMANN, 1987a. Personal Communication. Purdue University, West Lafayette, IN 47907, Computer Science Dept.

M. KALKBRENER, 1987. *Solving Systems of Algebraic Equations by Using Gröbner Bases.* Proc. of the EUROCAL '87 Conference, Leipzig, 2–5 June 1987, to appear.

D. KAPUR, 1986. *A Refutational Approach to Geometry Theorem Proving.* In: Proceedings of the Workshop on Geometric Reasoning, Oxford University, June 30 - July 3, 1986, to appear in *Artificial Intelligence.*

D. KAPUR, 1987. *Algebraic Reasoning for Object Construction from Ideal Images.* Lecture Notes, Summer Program on Robotics: Computational Issues in Geometry, August 24–28, Institute for Mathematics and its Applications, Univ. of Minneapolis.

A. KANDRI-RODY, 1984. *Effective Methods in the Theory of Polynomial Ideals.* Ph. D. Thesis, Rensselaer Polytechnic Institute, Troy, New York, Dept. of Computer Science.

H. KREDEL, 1987. *Primary Ideal Decomposition.* Proc of the EUROCAL '87 Conference, Leipzig, 2–5 June 1987, to appear.

B. KUTZLER, 1987. *Implementation of a Geometry Proving Package in SCRATCH-PAD II.* Proceedings of the EUROCAL '87 Conferenc, Leipzig, 2–5 June, 1987, to appear.

B. KUTZLER, S. STIFTER, 1986. *On the Application of Buchberger's Algorithm to Automated Geometry Theorem Proving.* J. of Symbolic Computation, 2/4, 389–398.

D. LAZARD, 1985. *Ideal Bases and Primary Decomposition: Case of Two Variables.* J. of Symbolic Computation 1/3, 261–270.

R. P. PAUL, 1981. " Robot Manipulators: Mathematics, Programming, and Control". The MIT Press, Cambridge (Mass.), London.

F. P. PREPARATA, M. I. SHAMOS, 1985. "Computational Geometry". Springer-Verlag, New York, Berlin, Heidelberg.

T. W. SEDERBERG, D. C. ANDERSON, 1984. *Implicit Representation of Parametric Curves and Surfaces.* Computer Vision, Graphics, and Image Processing 28, 72–84.

D. SPEAR, 1977. *A Constructive Approach to Ring Theory.* Proc. of the MAC-SYMA Users' Conference, Berkeley, July 1977 (R. J. Fateman ed.), The MIT Press, 369–376.

B. STURMFELS, 1987. Private Communication. Institute for Mathematics and its Applications.

W. TRINKS, 1978. *On B. Buchberger's Method for Solving Systems of Algebraic Equations (German).* J. of Number Theory 10/4, 475–488.

A. VAN DEN ESSEN, 1986. *A Criterion to Decide if a Polynomial Map is Invertible and to Compute the Inverse.* Report 8653, Catholic University Nijmegen (The Netherlands), Dept. of Mathematics.

B. L. VAN DER WAERDEN, 1953. "Modern Algebra I, II", Frederick Ungar Publ. Comp., New York.

F. WINKLER, 1986. *Solution of Equations I: Polynomial Ideals and Gröbner Bases.* Proc. of the Conference on Computers and Mathematics, Stanford University, July 30 - August 1, 1986, to appear.

W. T. WU , 1978. *On the Decision Problem and the Mechanization of Theorem Proving in Elementary Geometry.* Scientia Sinica 21, 150-172.

# GEOMETRY IN DESIGN:
# THE BEZIER METHOD

GERALD FARIN*

Abstract: *The Bezier method for the representation of polynomial curves and surfaces is outlined, with emphasis on a geometric viewpoint. Several examples are given to underline the usefulness of the geometric approach to curve and surface design.*

**1. Introduction and history.** In the late fifties computer technology had reached a point where it was possible to drive a machining tool in order to manufacture stamps and dies – in much the same fashion that a pen plotter is driven to produce a drawing. In order to make use of this possibility, the stamp or die surface had to be represented in a computer-compatible form. This is not difficult for conventional surfaces such as spheres, planes, cylinders etc.; no method was known, however, to achieve that goal for general, so-called "sculptured" surfaces.

In 1958, P. de Casteljau of Citroen/Paris, having just received his Ph.D. in Mathematics, tackled and solved this surface representation problem. He called his method "surfaces à pôles", and Citroen employed it for the design and manufacture of its automobiles. However, they were extremely secretive about any details of the method; thus de Casteljau's work was never published.

A little later, Renault/Paris, also realizing the importance of the surface representation problem, and having heard about developments at Citroen, began development of their own system. These efforts were led by P. Bezier, an engineer who had been in the design and manufacturing business since the thirties. The result of these efforts was the CAD system UNISURF.

The approach taken by Bezier was completely different from that of de Casteljau, yet

* Arizona State University, Tempe, AZ 85287. This work was supported in part by Department of Energy contract DE–AC02-85ER12046 and by NSF grant DCR-8502858.

the underlying mathematics for both systems is the same: the use of Bernstein polynomials for the representation of curves and surfaces. While de Casteljau made explicit use of this representation, it took until 1972 when R. Forrest realized that Bernstein polynomials were also the natural tool to describe the curves and surfaces of the UNISURF system.

In 1946, I. Schoenberg developed the theory of B-splines; in 1974, W. Gordon and R. Riesenfeld, knowing about Bezier curves, reformulated B-splines so that they could be used in a Bezier-like fashion. Both the B-spline and the Bezier curve and surface description use the same concept: a curve or surface is defined by a discrete set of *control points*.

The use of control points allowed a separation of the mathematics underlying a design system from the actual use of it. Designers, trained in descriptive geometry but not in calculus, can easily get a feeling of how to design a curve by adjusting control points – Bezier or B-spline control points alike.

In this article, we shall see how the use of control point schemes also facilitates the underlying mathematics.

**2. Bezier curves – a brief introduction.** We follow here the original approach taken by de Casteljau (de Casteljau 1963) – it is mathematically more fruitful than Bezier's own approach.

Let a polygon of 3D points $\mathbf{b}_0, \cdots, \mathbf{b}_n$ be given. The following algorithm generates a point on a curve – the *Bezier curve* – that is determined by the $\mathbf{b}_i$, the *control polygon*:

For $r = 1, \cdots, n$ and $i = 0, \cdots, n - r$ and a parameter value $t$, generate points

$$(1) \qquad \mathbf{b}_i^r(t) = (1 - t)\mathbf{b}_i^{r-1}(t) + \mathbf{b}_{i+1}^{r-1}(t),$$

where $\mathbf{b}_i^0 = \mathbf{b}_i$ and $\mathbf{x}(t) = \mathbf{b}_0^n(t)$ is the desired point on the curve. The construction is illustrated in figure 1.

The above construction is the original *de Casteljau algorithm*. Its underlying geometry can also be used to derive several important properties of Bezier curves:

*Affine invariance:* Let $\mathcal{B}$ be the (linear) operator $\mathcal{B}[\mathbf{b}_0, \cdots, \mathbf{b}_n; t] = \mathbf{x}(t)$. Let $\Phi$ be an affine map. Then $\Phi\mathcal{B} = \mathcal{B}\Phi$. This follows from the fact that the de Casteljau algorithm

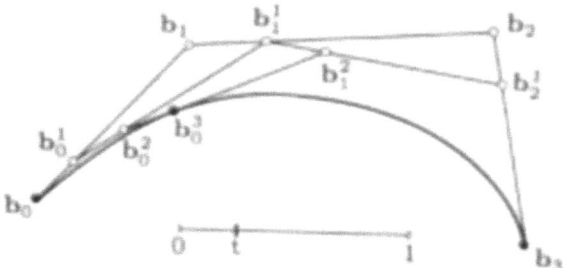

Figure 1: The de Casteljau construction

uses piecewise linear interpolation only, which is affinely invariant.

*Endpoint interpolation:* The Bezier curve passes through $\mathbf{b}_0$ and $\mathbf{b}_n$. This follows immediately from the construction.

*Convex hull property:* For $t \in [0, 1]$, the curve lies in the convex hull of the polygon. This follows since the de Casteljau algorithm uses *convex combinations* only.

A simple inductive proof shows that the curve generated by (1) can explicitly be written as

$$(2) \qquad \mathbf{x}(t) = \sum_{i=0}^{n} \mathbf{b}_i B_i^n(t),$$

where $B_i^n(t) = \binom{n}{i} t^i (1-t)^{n-i}$ are called *Bernstein polynomials*. The recursive construction (1) is mirrored by a recursion that the Bernstein polynomials satisfy:

$$B_i^r(t) = (1 - t) B_i^{r-1}(t) + t B_{i-1}^{r-1}(t).$$

Two other important properties are *positivity:* $B_i^n(t) \geq 0$ for $t \in [0, 1]$ and *partition of unity:* $\sum B_i^n(t) \equiv 1$.

All these properties may be proven analytically; however, they are also a consequence of the construction (1) and thus admit more geometric proofs.

For the derivatives of a Bezier curve we find

(3)
$$\frac{d^r}{dt^r}\mathbf{x}(t) = \frac{n!}{(n-r)!}\sum_{i=0}^{n-r}\Delta^r\mathbf{b}_i B_i^{n-r}(t),$$

where $\Delta^r$ is the iterated forward difference operator. Thus one *differentiates* a Bezier curve by *differencing* its control points. In particular, one sees that the curve is tangent to $\Delta\mathbf{b}_0$ and $\Delta\mathbf{b}_{n-1}$, a fact that will be useful when one considers composite curves.

As one final item, we discuss the *subdivision* of a Bezier curve. As the parameter value $t$ divides the interval $[0,1]$ into a right and a left subinterval, the point $\mathbf{x}(t)$ divides the curve into two pieces. Looking at figure 1, one gets the impresssion that some of the intermediate $\mathbf{b}_i^r$ provide control polygons for these subarcs of the original curve. This is indeed true: the arc between $\mathbf{b}_0$ and $\mathbf{x}(t)$ has control points $\mathbf{b}_0^i; i = 0, \cdots, n$ and the arc between $\mathbf{x}(t)$ and $\mathbf{b}_n$ has control points $\mathbf{b}_i^{n-i}; i = 0, \cdots, n$.

As a consequence, the de Casteljau algorithm not only generates a point on the curve, but also the tangent there:
$$\frac{d}{dt}\mathbf{x}(t) = n(\mathbf{b}_1^{n-1} - \mathbf{b}_0^{n-1}).$$
Analogous results hold for higher derivatives.

We will now show how these elementary curve properties can be used to formulate important generalizations and applications.

**3. Cubic spline curves.** A *spline curve* $\mathbf{s}(u)$ is composed of several cubic pieces so as to form a continuous space curve that is typically twice differentiable. Each piece of the spline curve is defined over an interval $[u_i, u_{i+1}]$ of a partition of the real line and has Bezier points $\mathbf{b}_{3i}, \cdots, \mathbf{b}_{3i+3}$.

Let us now consider two adjacent spline segments, defined over $[u_{i-1}, u_i]$ and $[u_i, u_{i+1}]$, respectively. Both pieces are $r$ times continuously differentiable if

(4)
$$\mathbf{b}_{3i+\rho} = \mathbf{b}_{3i-\rho}^\rho(\frac{\Delta u_{i-1} + \Delta u_i}{\Delta u_{i-1}}); \quad \rho = 0, \cdots, r.$$

This condition is easily proved from the subdivision procedure outlined above and the derivative formula (3). The geometric interpretation is as follows:

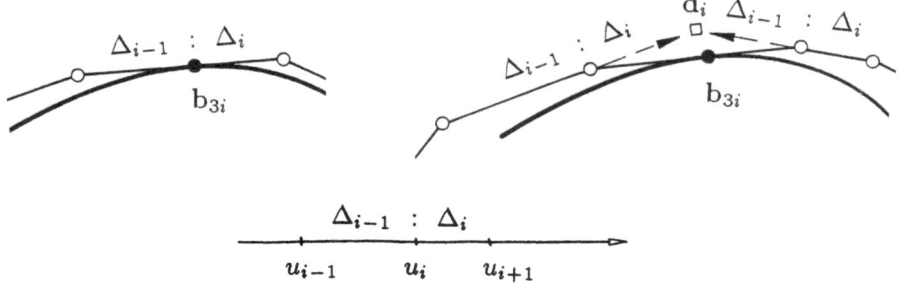

**Figure 2:** The $C^1$ and $C^2$ conditions.

The curve is once differentiable at $u_i$ if a) $b_{3i-1}, b_{3i}, b_{3i+1}$ are collinear and b) $b_{3i}$ divides $b_{3i-1}$ and $b_{3i+1}$ in the ratio $\Delta u_i : \Delta u_{i+1}$, see figure 2.

The curve is twice differentiable if an auxiliary point $d_i$ exists such that $b_{3i-1}$ divides $b_{3i-2}$ and $d_i$ in the ratio $\Delta u_{i-1} : \Delta u_i$ and also such that $b_{3i+1}$ divides $d_i$ and $b_{3i+2}$ in the same ratio, see figure 2. This condition was first derived by W. Boehm (Boehm 1977).

In order to check if the spline curve s is twice differentiable, one therefore constructs the auxiliary point $d_i$ from the left and from the right segment; if both constructions yield the same point, the curve is twice differentiable.

The auxiliary points $d_i$ play a far more important role than just being "check points": for a twice differentiable curve s, we can construct all points $d_i$. The polygon that is formed by these points is related to the curve in a one-to-one manner, just as is the piecewise Bezier polygon. However, while the piecewise Bezier polygon of a $C^2$ curve contains redundant information, the polygon of the $d_i$ carries no redundancy. It is called the *B-spline polygon* of the spline curve $s(u)$.

We can write all Bezier points as linear combinations of the $d_i$. The necessary coefficients can be read off from figure 3.

We have developed the notion of a B-spline curve in a very elementary and geometric manner. Other derivations (see e.g. Bartels *et al* 1987) require a more abstract mathematical analysis which does not give the geometric insight provided by our derivation.

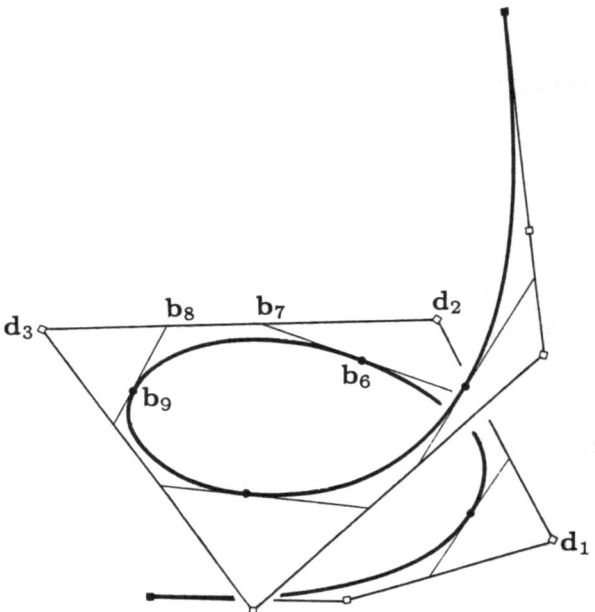

**Figure 3:** B-spline polygon with corresponding piecewise Bezier polygon.

**4. Curvature.** The curvature $\kappa(t)$ of a curve $\mathbf{x}(t)$ is given by

$$\kappa(t) = \frac{||\dot{\mathbf{x}} \times \ddot{\mathbf{x}}||}{||\dot{\mathbf{x}}||^3}.$$

We are here interested in piecewise cubic Bezier curves, more specifically in their curvature at the point $\mathbf{b}_{3i}$, and we therefore insert the relevant derivatives from (3):

$$\kappa(u_i^+) = \frac{2}{3} \frac{||\Delta \mathbf{b}_{3i} \times \Delta^2 \mathbf{b}_{3i}||}{||\Delta \mathbf{b}_{3i}||^3}.$$

The term $u_i^+$ is used to indicate that our attention is limited to the interval $[u_i, u_{i+1}]$.

With the notation from figure 4, we can rewrite this as

$$\kappa(u_i^+) = \frac{2}{3} \frac{h_i^+}{||\Delta \mathbf{b}_{3i}||^2}.$$

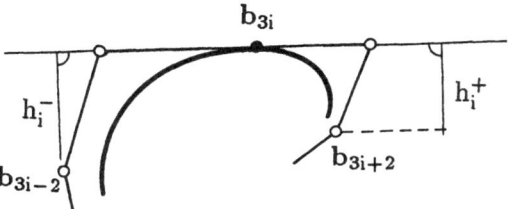

**Figure 4:** Two adjacent segments with curvature continuity.

Therefore the two curve segments defined over $[u_{i-1}, u_{i+1}]$ have continuous curvature at $u_i$ if

(5)
$$\frac{h_i^-}{||\Delta \mathbf{b}_{3i-1}||^2} = \frac{h_i^+}{||\Delta \mathbf{b}_{3i}||^2}.$$

If, in addition, we also require that $\mathbf{b}_{3i-2}, \cdots, \mathbf{b}_{3i+2}$ lie in one plane $P$, we have ensured continuity of the osculating plane between the two curve segments: both will then have $P$ as their osculating plane at $\mathbf{b}_{3i}$.

Curvature continuous splines will in general fail the $C^2$ test from the preceding section: instead of one auxiliary point $\mathbf{d}_i$, there will in general be two of them, $\mathbf{d}_i^-$ and $\mathbf{d}_i^+$, say. The conditions for curvature and osculating plane continuity can now be unified: two adjacent curve segments share the same curvature and osculating plane at $u_i$ if $\mathbf{d}_i^+ - \mathbf{d}_i^- = \nu(\mathbf{b}_{3i+1} - \mathbf{b}_{3i-1})$ for some $\nu \in \mathbf{R}$. (This is, by the way, the same $\nu$ as used by Nielson in the derivation of $\nu$-splines, see Nielson(1974) and Farin(1985).)

Splines with continuity of curvature and osculating plane are sometimes called $V^2$ splines (from *visually continuous*). We can now generalize the algorithm for the generation of the Bezier points of a $C^2$ curve to a $V^2$ curve (Farin 1982): instead of *computing* Bezier points on the poygon legs $\mathbf{d}_i, \mathbf{d}_{i+1}$, we can *choose* them arbitrarily. The remaining points $\mathbf{b}_{3i}$ are then computed from (5).

This algorithm is entirely geometric: it relies only on the geometry of the control polygons and does not need the notion of the knot sequence $u_i$ any more. This frees bcth designer and system programmer from the usually artificial determination of the knots $u_i$. Two other methods – Boehm's $\gamma$-splines (Boehm(1987)) and Barsky's $\beta$-splines (Bartels, Beatty,

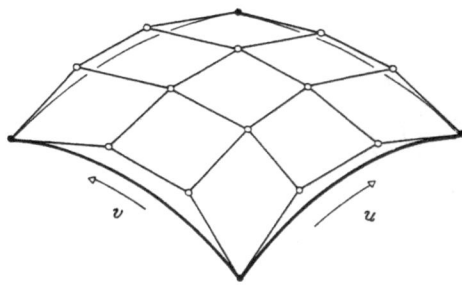

**Figure 5:** A rectangular Bezier surface and defining control net.

Barky 1987) – require the knot sequence to be determined (called $\beta_i^{(1)}$ instead of $\Delta u_i$ by Barsky).

$V^2$ splines are intrinsically more complicated than $C^2$ splines: they do not form a linear space, while $C^2$ splines do. This precludes the use of $V^2$ splines in tensor product methods, as the resulting surfaces would not be curvature continuous.

**5. Rectangular patches.** A rectangular Bezier surface (or patch) is defined by

$$\mathbf{x}(u,v) = \sum_{i=0}^{m}\sum_{j=0}^{n} \mathbf{b}_{i,j} B_i^m(u) B_j^n(v).$$

The $\mathbf{b}_{i,j}$ form a *control net*, as shown in figure 5. Most of its properties are easily derived from the univariate case, for instance we again have the convex hull property and affine invariance. The "endpoint interpolation" property becomes: The boundary curves of the patch have the boundaries of the control net as control polygons.

Another important patch type that is used in CAD is the so-called *Coons patch*. It constructs a surface from any four boundary curves. Let the final surface be $\mathbf{x}(u,v)$. Then its (given) boundaries are $\mathbf{x}(0,v), \mathbf{x}(u,1)$ etc. The Coons formula for the final surface is:

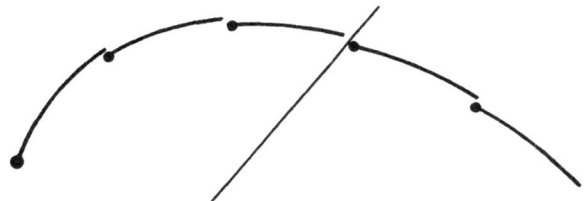

**Figure 6:** "Numerical holes" in a piecewise monomial curve.

$$\mathbf{x}(u,v) = (1-u)\mathbf{x}(0,v) + u\mathbf{x}(1,v) + (1-v)\mathbf{x}(u,0) + v\mathbf{x}(u,1)$$
$$- (1-u \quad u) \begin{pmatrix} \mathbf{x}(0,0) & \mathbf{x}(0,1) \\ \mathbf{x}(1,0) & \mathbf{x}(1,1) \end{pmatrix} \begin{pmatrix} 1-v \\ v \end{pmatrix}.$$

If the four boundary curves are Bezier curves, it is clear that the Coons patch is a tensor product Bezier patch. What is the Bezier net of this surface? The answer to this question is surprisingly simple: interpret the polygons of the boundary curves as piecewise linear curves. Apply the Coons formula to them and obtain a piecewise bilinear surface. This surface is the desired Bezier net.

This result also holds for B-spline surfaces (Farin 1987).

**6. Stability.** When working in the automotive industry, I was involved in writing curve/plane intersection routines. Our curves, piecewise cubics, were stored in the piecewise monomial form: $\mathbf{s}(u) = \mathbf{a}_0 + (u - u_i)\mathbf{a}_1 + (u - u_i)^2\mathbf{a}_2 + (u - u_i)^3\mathbf{a}_3$. These piecewise monomials can be viewed as Taylor expansions of $\mathbf{s}$ at the points $\mathbf{s}(u_i)$: they are very accurate for $u-$ values slightly larger than $u_i$, but they deteriorate further away from $u_i$. Consequently, our spline curves had "numerical holes", as is shown somewhat exaggerated in figure 6.

This situation was discovered when a user complained about a failure of the spline/plane intersector: he encountered a case where the plane passed exactly through a numerical hole. An adjustment of tolerances produced the desired intersection – however, the problem of ill-defined piecewise monomial curves remains.

It is easy to see, of course, that the problem could not have arisen had the curve been represented in the piecewise Bezier form: since $\mathbf{b}_{3i} = \mathbf{s}(u_i)$ is stored only once, right and left limits agree.

Recently R. Farouki and V. Rajan (1987) addressed an interesting property of Bezier curves: consider the functional case $f(x) = b_0 B_0^n(x) + b_1 B_1^n(x) + \cdots$. The polynomial $f$ has certain roots in the interval $[0, 1]$. If we perturb the coefficients $b_i$ by small amounts $\epsilon_i$, the location of the roots will change. For a specific root $x_j$ the *condition number* of that root with respect to the coefficients $b_i$ is defined as the rate of displacement of $x_j$ relative to the perturbation of the $b_i$.

Farouki and Rajan show that all roots of any polynomial are better conditoned if that polynomial is expressed in Bezier form rather than in the monomial form.

**7. Conclusions.** We have given several examples of the usefulness of the geometric approach to curve and surface design that is provided by the Bezier (or B-spline) form. Instead of elaborating on this, we conclude with two additional remarks:

As an associate editor of the journal CAGD, I have seen the following happen several times: papers were submitted that used the piecewise monomial or the piecewise Hermite form for the development of a new curve or surface scheme. In many cases, this gave rise to a lengthy development of the material – prompting referees to require the Bezier (or B-spline) form to be used. In almost all such cases, this shortened and clarified those papers considerably.

Finally: another type of Bezier surface exists that was not mentioned here, the so-called Bezier triangles. Those are patches that are defined over a triangular instead of a rectangular domain (Farin 1986, de Boor 1987). These patches can be used to analyze piecewise polynomial functions over a triangulation of the plane. The Bezier form for these piecewise polynomials has proved to be so successful that practically no other representation is used in current publications (see e.g. Alfeld 1987).

## REFERENCES

P. Alfeld (1987): A case study of multivariate piecewise polynomials, in: *Geometric Modeling*, G. Farin (ed.), SIAM, 149-159.

C. de Boor (1987): B-form basics, in: *Geometric Modeling*, G. Farin (ed.), SIAM, 131-148.

R. Bartels, J. Beatty, B. Barsky (1987): *An Introduction to the Use of Splines in Computer Graphics*, Morgan-Kaufmann.

W. Boehm (1977): Cubic B-spline curves and surfaces in CAGD, Computing 19, 29-34.

W. Boehm(1987): Smooth curves and surfces, in: *Geometric Modeling*, G. Farin (ed.), SIAM, 175-184.

W. Boehm, G. Farin, J. Kahmann (1984):A survey of curve and surface methods in CAGD, Computer Aided Geometric Design 1, 1-60.

P. de Casteljau (1963): Courbes et surfaces à pôles; André Citroen, Paris.

G. Farin (1982): Visually $C^2$ cubic splines, Computer Aided Design 14, 137-139.

G. Farin (1986): Triangular Bernstein-Bezier patches, CAGD 3, 83-128.

G. Farin (1987): The commutativity of tensor product and Boolean surface schemes. Being revised.

R. Farouki and V. Rajan (1987): On the numerical condition of Bernstein polynomials, to appear in CAGD.

R. Forrest (1972): Interactive interpolation and approximation by Bezier polynomials, The Computer Journal 15, 71-79.

W. Gordon and R. Riesenfeld (1974): B-spline curves and surfaces, in: *Computer Aided Geometric Design*, R. Barnhill and R. Riesenfeld (eds.), Academic Press, 95-126.

G. Nielson (1974): Some piecewise polynomial alternatives to splines under tesnion, in: *Computer Aided Geometric Design*, R Barnhill and R. Riesenfeld (eds.), Academic Press, 209-235.

I. Schoenberg (1946): Contributions to the problem of approximation of equidistant data by analytic functions, Quart. Appl. Math 4, 45-99.

# ALGEBRAIC CURVES

## CHRISTOPH M. HOFFMANN[1]

**Abstract.** We consider the problem of tracing algebraic curves by computer, using a numerical technique augmented by symbolic computations. In particular, all singularities are analyzed correctly. The methods presented find application in solid modeling and robotics.

**Key Words.** Algebraic curves, surface intersection, desingularisation,

**1. Introduction.** In this paper we discuss preliminary results for tracing algebraic curves. Planar algebraic curves of the form $f(x, y) = 0$ are considered, as are space curves that are the intersection of two algebraic surfaces, $f(x, y, z) = 0$ and $g(x, y, z) = 0$. The scenario is as follows: We are given a point $p$ on the curve, and a direction of traversal. We wish to trace succeeding curve points on the same branch, and we would like to trace them through singularities.

A reliable solution to this problem has immediate applications to solid modeling and geometric design. For example, the well-known Boolean operations on solids require tracing surface intersections, for the purpose of determining the surface of the resulting solid [9]. Here, algebraic space curves arise when the intersecting solids are bounded by algebraic faces. If the surface is composed of parametric patches, e.g., rational B-spline surfaces, then planar algebraic curves can be obtained [6].

Since an extensive amount of curve tracing is required for each modeling operation, it is advisable to pay attention to efficiency as well as reliability. For this reason, precise methods such as the cylindrical algebraic decomposition due to Collins and its variants [4,11] have not been considered here. This does not imply that these presently very compute intensive methods must remain of theoretical interest only, but significantly more work is required before we can accurately gauge whether in the specialized context of solid modeling versions of these algorithms exist that can be used without serious efficiency degradations.

As point of departure, we use a straightforward numerical method that approximates the curve locally by its truncated Taylor series, and then performs a Newton iteration to correct the accumulated error. This approximation has many interesting properties. For instance, the Taylor series is a special case of the Puiseux series that can be used to approximate the curve at singular points as well as at regular points. Thus, there is a basis for developing a uniform framework for studying approximations to algebraic curves. Note, however, that there are alternative curve approximations based on special classes of polynomials that offer different

---

[1]Computer Science Department, Purdue University. Work supported in part by the Office of Naval Research under contract N0014-86-K-0465 and the Institute for Mathematics and its Applications at the University of Minnesota.

advantages, [7], and more study is needed before the relative merits can be fully appreciated.

In many cases the numerical procedure suffices and copes acceptably well with certain singularities, e.g., with normal crossings and with tacnodes that are not very complicated. It is not fully reliable, however, and will fail at cuspidal singularities. For this reason it is augmented by a mapping technique that exploits the fact that by a suitable birational map any singularity can be resolved. That is, such a map will transform the singular point into one or more nonsingular ones, while not creating new singularities. Fortunately, suitable maps can be found easily in the planar case. In the space curve case the situation is not so simple, and more work is required to find attractive algorithms.

Even in the planar case a number of details must be addressed before curve desingularization can be automated. These include finding reliably the locus of the singularity, controlling numerical inaccuracies that arise from the various desingularization maps, and establishing the correct correspondence of orientation between the curve and its transformations.

We structure this paper as follows: After explaining concepts and notation in Section 2, we devote Section 3 to a description of the simple numerical procedure for following curves. In Section 4, we explain the correspondence between the Taylor series and the notion of a place of a curve, which yields a straightforward method for extending the numerical procedure so as to cope with low order singularities consistently. The extension is of limited value, however, as it involves solving systems of polynomial equations whose degree depends on the order of the singularity analyzed. Section 5 concentrates on desingularizing planar algebraic curves and discusses how problems such as orientation correspondence can be solved. Examples of various types of singularities are also given. Section 6 then discusses the desingularization of space curves. Sections 3, 4 and 5 summarize work reported in [2,8].

**2. Concepts and Notation.** We consider algebraic space curves given as the intersection of two algebraic surfaces $f(x, y, z) = 0$ and $g(x, y, z) = 0$, where $f$ and $g$ are polynomials in $x$, $y$, and $z$. This is not the most general definition of space curves; certain curves require intersecting more than two surfaces in order to exclude extraneous components. The partial derivatives of $f$ are written by subscripting, e.g., $f_x = \partial f/\partial x$, $f_{xy} = \partial^2 f/(\partial x \partial y)$, and so on. Since $f$ and $g$ are analytic, $f_{xy} = f_{yx}$ etc.

Vectors and vector functions are denoted by bold letters. The *inner product* of two vectors a and b is the scalar $\mathbf{a} \cdot \mathbf{b}$. The *length* of the vector a is $|\mathbf{a}| = \sqrt{\mathbf{a} \cdot \mathbf{a}}$. The *cross product* of the vectors is the vector $\mathbf{a} \times \mathbf{b}$.

The *gradient* of the surface $f$ at the point $p = (x, y, z)$ is the vector $\nabla f = (f_x, f_y, f_z)$, where the partials are evaluated at $p$. If not zero, it is a vector that is normal to the surface at $p$. A point $p = (x, y, z)$ is *regular* on $f$ if the gradient of $f$ at $p$ is not null; otherwise the point is *singular*. The *Hessian* of $f$ at point $p$ is the tensor

$$H_f = \begin{pmatrix} f_{xx} & f_{xy} & f_{xz} \\ f_{yx} & f_{yy} & f_{yz} \\ f_{zx} & f_{zy} & f_{zz} \end{pmatrix}$$

where the partials are evaluated at $p$.

The intersection curve of $f$ and $g$ is denoted $r(s)$ and is considered a vector function of the argument $s$. In Section 3, we will determine an approximation of $r$ in which $s$ is the arc length measured from some initial point on the curve. As usual, the derivatives of $r(s)$ are denoted $r'$, $r'' \ldots r^{(m)}$.

A point $p$ of the intersection curve $r(s)$ is *regular* if $p$ is regular on both $f$ and $g$ and if the gradients $\nabla f$ and $\nabla g$ are linearly independent. That is, the surfaces are not singular at $p$ and intersect transversally. A point $p$ is *singular* on $r(s)$ for one of the following reasons:

1. The gradients $\nabla f$ and $\nabla g$ are nonzero and linearly dependent.

2. One of the gradients, say $\nabla g$ is zero, but the other is not.

3. The gradients $\nabla f$ and $\nabla g$ are both zero.

We note that Cases 1 and 2 do not differ in substance.

The *initial form* of $f$ is the polynomial formed by all terms of lowest order in $f$. For example, the initial form of $x^2 - 2x + y^2 + z^2$ is $-2x$. The initial form approximates the surface in the neighborhood of the origin. In the above example, the surface "looks like" the plane $x = 0$ near the origin. This plane is the tangent plane to $f$ at the origin. In particular, if $p$ is the origin, then $p$ is regular on $f$ if the initial form of $f$ is linear. Otherwise $p$ is singular.

At each point $p = r(s)$ of a space curve, an intrinsic coordinate system is provided by the *orthonormal triad*. The triad consists of three perpendicular unit vectors, namely the tangent $t$, the principal normal $h$, and the binormal $b$. The Frenet-Serret formulas relate the triad to arc length, curvature, and torsion of the curve at $p$. With $s$ the arc length, $\rho$ the radius of curvature, and $\tau$ the radius of torsion we have

$$\frac{dt}{ds} = \frac{1}{\rho}h, \qquad \frac{db}{ds} = -\frac{1}{\tau}h, \qquad \frac{dh}{ds} = \frac{1}{\tau}b - \frac{1}{\rho}t.$$

A planar algebraic curve is given by its implicit equation $f(x,y) = 0$, $f$ a polynomial in $x$ and $y$. Equivalently, we can think of the curve as the intersection of the surfaces $f$ and $z = 0$. Note that $f$ is then a cylinder with the base line $f(x,y) = 0$. All concepts explained above can therefore be transferred to the planar case.

**3. Numerical Tracing.** For space curves, the simplest situation arises when tracing the curve in a neighborhood in which both surfaces are nonsingular and intersect each other transversally. This means that the gradients of $f$ and $g$ do not vanish along the curve and are linearly independent vectors. In this case we formulate a system of linear equations from which to obtain the local approximation to the intersection curve at the point $p$.

**3.1. The Basic Method for Space Curves.** Since the curve $\mathbf{r}(s)$ must satisfy both $f$ and $g$ identically, each coefficient in the Taylor series of $f(\mathbf{r}(s))$ and $g(\mathbf{r}(s))$ will be zero. Let $p = \mathbf{r}(0)$ be a point on the intersection. Then

$$f(\mathbf{r}(s)) = f(0) + s\nabla f \cdot \mathbf{r}'(0) + \frac{s^2}{2}[\nabla f \cdot \mathbf{r}''(0) + \mathbf{r}'(0) \cdot H_f \cdot \mathbf{r}'(0)] + \cdots$$

and similarly,

$$g(\mathbf{r}(s)) = g(0) + s\nabla g \cdot \mathbf{r}'(0) + \frac{s^2}{2}[\nabla g \cdot \mathbf{r}''(0) + \mathbf{r}'(0) \cdot H_g \cdot \mathbf{r}'(0)] + \cdots$$

This leads to the following system of equations, for $m = 1, 2, \ldots$:

(1)
$$\begin{aligned} \nabla f(p) \cdot \mathbf{r}^{(m)}(0) &= B_{1,m} \\ \nabla g(p) \cdot \mathbf{r}^{(m)}(0) &= B_{2,m} \end{aligned}$$

The quantities $B_{1,m}$ and $B_{2,m}$ are expressions in the partial derivatives of $f$ and $g$ and the lower-order derivatives of $\mathbf{r}$. For $m = 1$ we have

$$B_{1,1} = B_{2,1} = 0$$

and for $m = 2$

$$\begin{aligned} B_{1,2} &= -\mathbf{r}' \cdot H_f \cdot \mathbf{r}' \\ B_{2,2} &= -\mathbf{r}' \cdot H_g \cdot \mathbf{r}' \end{aligned}$$

For higher values of $m$ the expressions $B_{i,m}$ are more complex.

With the assumption of independent gradients at $\mathbf{r}(0)$, the solution to the system has the form

$$\alpha_m \nabla f + \beta_m \nabla g + \gamma_m \nabla f \times \nabla g$$

$\gamma_m$ can be chosen arbitrarily, and the other coefficients satisfy the nonsingular system

(2)
$$\begin{pmatrix} \nabla f \cdot \nabla f & \nabla f \cdot \nabla g \\ \nabla g \cdot \nabla f & \nabla g \cdot \nabla g \end{pmatrix} \begin{pmatrix} \alpha_m \\ \beta_m \end{pmatrix} = \begin{pmatrix} B_{1,m} \\ B_{2,m} \end{pmatrix}$$

System (1) is underdetermined. When using it to approximate the intersection curve, the choices of $\gamma_m$ relate to the ability to obtain equivalent approximations, e.g., parameterized by $cs$ instead of $s$, where $c \neq 0$ is a constant. By using the Frenet-Serret formulas, we can interpret these choices. We set $\mathbf{r}' = d\mathbf{r}/ds = \mathbf{t}$, to obtain

$$\mathbf{r}'' = \frac{1}{\rho}\mathbf{h}, \qquad \mathbf{r}''' = \frac{1}{\rho\tau}\mathbf{b} - \frac{1}{\rho^2}\mathbf{t} - \frac{d\rho/ds}{\rho^2}\mathbf{h}$$

For $m = 1$ we have $B_{1,1} = B_{2,1} = 0$, hence $\alpha_1 = \beta_1 = 0$. We let

$$\mathbf{r}' = \frac{\nabla f \times \nabla g}{|\nabla f \times \nabla g|}$$

be the solution of the system, so that the parameter $s$ corresponds to the arc length of the intersection curve.

For $m = 2$ we choose $\gamma_2 = 0$. This implies that $\mathbf{r}''$ is orthogonal to $\mathbf{r}'$, and that from the system solution $\mathbf{r}'' = \alpha_2 \nabla f + \beta_2 \nabla g = \mathbf{h}/\rho$ both the principal normal $\mathbf{h}$ and the radius of curvature $\rho$ is determined.

For $m = 3$ we get $\gamma_3 = -(\mathbf{r}'' \cdot \mathbf{r}'')$, since both $\mathbf{b}$ and $\mathbf{h}$ are orthogonal to $\mathbf{t}$. This determines both $\mathbf{r}'''$ and $\tau$.

The curve approximation so determined can be used in a neighborhood of $\mathbf{r}(0)$ whose size can be estimated by the magnitude of the second and third order terms. If the ratio of $\delta^3 |\mathbf{r}'''|/6$ to $|\mathbf{r} + \delta \mathbf{r}' + \delta^2 \mathbf{r}''/2|$ is small, then the higher order derivatives contribute little and we have not deviated too much from the true intersection. By halving or doubling a standard distance repeatedly, the stepping size can be adjusted according to curvature and torsion. It is necessary to establish a minimum stepping size since near-singular curves can have areas of arbitrarily high curvature where repeated halving might lead to unacceptable running times.

At the end of the current approximation to $\mathbf{r}$ a point $P_0$ is reached that is near the curve of intersection but not on it. Beginning with this point, we find a sequence of points $P_1, P_2, \ldots$ that converges to a point $p$ on the curve, using Newton's method. With $P_{k+1} = P_k + \Delta_k$, we want to solve the system

$$\begin{aligned}
\nabla f(P_k) \cdot \Delta_k &= -f(P_k) \\
\nabla g(P_k) \cdot \Delta_k &= -g(P_k)
\end{aligned}$$
(3)

to obtain $\Delta_k \approx s P'_k$. Assuming linearly independent gradients,

$$P_k = \alpha_k \nabla f + \beta_k \nabla g + \gamma_k \mathbf{t},$$

where $\mathbf{t}$ is orthogonal to both gradients evaluated at $P_k$. Since a change in the direction of $\mathbf{t}$ does not change the values of $f$ and $g$ appreciably, we set $\gamma_k = 0$, thereby obtaining a unique solution for $\Delta_k$. We then set $P_{k+1} = P_k + \Delta_k$.

After the point $p$ is found with acceptable accuracy, a new approximation of $\mathbf{r}(s)$ centered at $p$ is determined.

**3.2. The Planar Case.** The planar curve $f(x, y) = 0$ arises as intersection of the cylinder $f(x, y) = 0$ and the plane $z = 0$. It can be traced this way as a space curve. In the system (1) the second equation specializes to $\mathbf{r}_z^{(m)} = 0$. Here $\mathbf{r}_z^{(m)}$ denotes the $z$ component of the $m^{th}$ derivative of $\mathbf{r}$. Moreover, since all partial derivatives by $z$ of $f$ are zero, the first equation takes the form

$$f_x \mathbf{r}_x^{(m)} + f_y \mathbf{r}_y^{(m)} = C_m$$

Thus, there is no difference between considering the intersection curve $\mathbf{r}$ or the planar curve.

As before, choosing $\gamma = 0$ for the Newton iteration means that we approach the curve along the local normal direction. An implementation could be specialized, but there appears to be no significant penalty for tracing the curve in space.

**3.3. Implementation.** The numerical tracing procedure has been implemented in Fortran by R. Lynch on a VAX 8600. With minor modifications, it has then been ported to a Symbolics Lisp machine. Figures 3.1 through 3.4 show some examples of curve traces so obtained. The planar curves have been traced as the intersection of $f(x, y) = 0$ with $z = 0$.

In our experience, nodal singularities cause no problems as long as the tangent directions of the intersecting branches are sufficiently separated. Since the curve orientation may reverse at singularities (c.f. Subsection 5.4 below), the tracing program must be augmented so as to maintain consistent tangent direction. However, the program cannot trace through cuspidal singularities. Many tacnodes are handled reliably, but inflections at the singularity are not recognized. Thus both the curve $f_1 = y^2 - x^4 - y^4 = 0$, shown in Figure 3.3, and the curve $f_2 = y^2 - x^6 - y^6 = 0$, shown in Figure 3.4, are traced as if they have two real components tangentially meeting at the origin. While this is correct for $f_1$, it is not correct for $f_2$ which consists of a single real component with the two branches at the origin each having a point of inflection.

**4. Algebraic Extensions of the Method.** We derived the equations (1) based on the assumption that the two surfaces intersect transversally and are not singular at the point $\mathbf{r}(0)$ of interest. Clearly, the radius of convergence of the power series about such a point cannot include any singular points of the curve. Nonetheless, the system of equations remains valid even when we are at a singular curve point. The reason for this is that Taylor's theorem is a special case of more general theorems.

Informally, a *place* of the planar curve $f(x, y) = 0$ is a pair of power series

$$x(s) = \sum_{k \geq 0} a_k s^k, \qquad y(s) = \sum_{k \geq 0} b_k s^k$$

such that $f(x(s), y(s))$ is identically zero. The *center* of a place is the point $(x(0), y(0))$ on the curve. Newton's Theorem states that centered at every curve point there is at least one place of $f$. Likewise, we define a place of the space curve $\mathbf{r}$ as the triple $(x(s), y(s), z(s))$. Since every space curve is the birational image of a planar curve [13,14], there is at least one place centered at every point of the space curve.

The connection between the notion of place and the equation system (1) in the previous section is established as follows. Centered at the point $p$ we consider the place

$$\mathbf{r}(s) = \sum_{k \geq 0} (a_k, b_k, c_k) s^k$$

where $p = \mathbf{r}(0)$. The derivative of the place is defined by

$$\mathbf{r}'(s) = \sum_{k \geq 1} (a_k, b_k, c_k) k s^{k-1}$$

Higher order derivatives are defined analogously.

Since $p$ is assumed to be on the intersection curve of $f$ and $g$, we know that $f(\mathbf{r}(s))$ and $g(\mathbf{r}(s))$ are identically zero, from which a system of equations is obtained for $m = 1, 2, 3, \ldots$

$$
\begin{aligned}
K_{1,m} &= 0 \\
K_{2,m} &= 0
\end{aligned}
$$

(4)

Here $K_{1,m}$ is the coefficient of $s^m$ in the power series $f(\mathbf{r}(s))$ and $K_{2,m}$ is the coefficient of $s^m$ in the power series $g(\mathbf{r}(s))$. This leads to the following

**Theorem.** For all $m \geq 1$ the equation $\nabla f \mathbf{r}^{(m)}(0) = B_{1,m}$ is equivalent to the equation $m! K_{1,m} = 0$.

The analogous statement holds for $g$ and $K_{2,m}$. The proof is by induction on the terms of $f$; see [8] for details. How the series are obtained from system (4) is illustrated by an example.

Consider the cylinders $f = x^2 + y^2 + 2x = 0$ and $g = x^2 + z^2 + 4x = 0$. Their intersection is an irreducible space curve of degree 4 with a singular point at the origin. At the singular point we have the following equations:

$$
\begin{aligned}
a_1 &= 0 \\
a_1 &= 0 \\
a_1^2 + 2a_2 + b_1^2 &= 0 \\
c_1^2 + 4a_2 + a_1^2 &= 0 \\
2a_1 a_2 + 2a_3 + 2b_1 b_2 &= 0 \\
2a_1 a_2 + 4a_3 + 2c_1 c_2 &= 0 \\
a_2^2 + 2a_1 a_3 + 2a_4 + b_2^2 + 2b_1 b_3 &= 0 \\
a_2^2 + 2a_1 a_3 + 4a_4 + c_2^2 + 2c_1 c_3 &= 0 \\
&\vdots
\end{aligned}
$$

One of the solutions to this system is

$$
\begin{aligned}
x(s) &= -s^2 \\
y(s) &= \sqrt{2}s - \frac{1}{2\sqrt{2}}s^3 \cdots \\
z(s) &= 2s - \frac{1}{4}s^3 \cdots
\end{aligned}
$$

In principle, this approach can be used to extend the tracing method of Section 3 so as to handle singular points, but it may become computationally expensive. Efficient strategies for solving these equations may exist. For example, one can always choose the coefficients of one series, say for $x(s)$, such that $|a_k| = 1$ for one specific $k$, and all other coefficients are zero [14].

5. **Planar Algebraic Curves.** The problem of tracing a curve reliably through any singularity is partially solved by the following theorem from algebraic geometry, e.g. [1]:

**Theorem.** *Any given algebraic plane curve can be transformed, by a birational transformation, into a curve devoid of singularities.*

The proof proceeds by an inductive argument that builds up the required birational transformation through a sequence of elementary, quadratic transformations. It is easy to understand that these transformations resolve ordinary singularities, but how progress is made on irregular singularities is more subtle, and we will not discuss it here. Different proofs of the theorem are found in, e.g., [1,10,13,14]. We restrict our attention to those transformation properties that are needed in order to understand how to derive an algorithm from the theorem.

**5.1. Desingularization.** We map a planar curve $f(x, y) = 0$ to a curve $g(x_1, y_1) = 0$ by the quadratic transformation

$$(5) \qquad\qquad x_1 = x \qquad y_1 = \frac{y}{x}$$

If $f$ has a singular point of order $m$ at the origin, then

$$f(x_1, x_1 y_1) = x_1^m g(x_1, y_1)$$

We call $f(x_1, x_1 y_1)$ the *total transform*, and $g(x_1, y_1)$ the *proper transform* of $f$. The $y_1$-axis is called the *exceptional line*. Figures 5.1 and 5.2 show two examples of a curve $f$ and its proper transform.

Intuitively speaking, applying the quadratic transformation separates intersecting curve branches that have different tangent directions. To appreciate this, note that the line $y - mx = 0$ is transformed to the line $y_1 - m = 0$. Moreover, all points $(x, y)$ of the $x$-$y$ plane with $x \neq 0$ are in $1 - 1$ correspondence with points $(x, y/x)$ of the $x_1$-$y_1$ plane. A point $(0, y)$ of the $x$-$y$ plane with $y \neq 0$ is mapped to infinity in the $x_1$-$y_1$ plane, and the origin of the $x$-$y$ is mapped to the exceptional line. The effect is that the singular point is "blown up" to the line $x_1 = 0$, and that the branches of $f$ at the origin are separated or, in a precise sense, made less singular. The proof of the theorem shows that after a finite number of quadratic transformations all singularities are removed. We wish to trace through a singular point as follows:

1. We trace the curve $f$ using the basic method of Section 3.

2. When approaching a singularity, we notice at some point $p$ that the determinant of the system becomes too small. We then locate the singular point as described below, and move it to the origin by translating the coordinate system.

3. Now the quadratic transformation is applied yielding the proper transform $g$.

4. We traverse $g$ beginning at the point $p_1$ corresponding to $p$, until we are past the singularity of $f$ and the system determinant is large enough to continue traversing $f$ accurately.

Note that we may have to traverse recursively iterated transforms of $f$, since the applied quadratic transform may not have fully desingularized the corresponding branch of $g$. Moreover, care must be exercised in correlating the orientation of $g$ and of $f$ to maintain proper traversal direction.

**5.2. Locating the Singularity.** When approaching a singular point $p$, the partial derivatives $f_x$ and $f_y$ of $f$ vanish. If the singularity has higher order, then higher order partial derivatives also vanish.

The singular point is defined as the intersection of the curves $f = 0$, $f_x = 0$, and $f_y = 0$. When traversing the curve $f$, we have approached the singularity to a point $P_0$ for which the partial $f_x$ and $f_y$ drop in value below a threshold $\mu$. We use a Newton iteration to construct a sequence of point approximations $P_i$ that converges to the singularity. The iteration is governed by the following system:

$$\begin{pmatrix} f_x(P_i) & f_y(P_i) \\ f_{xx}(P_i) & f_{xy}(P_i) \\ f_{xy}(P_i) & f_{yy}(P_i) \end{pmatrix} \begin{pmatrix} \delta_x \\ \delta_y \end{pmatrix} = - \begin{pmatrix} f(P_i) \\ f_x(P_i) \\ f_y(P_i) \end{pmatrix}$$

whose solution determines the next approximation to the intersection as $P_{i+1} = P_i + (\delta_x, \delta_y)$. Since this system is overconstrained, we solve it using the least-squares method by solving the $2 \times 2$ system

(6) $$A^T A \Delta = -A^T b$$

where $A$ is the $2 \times 3$ matrix, $\Delta = (\delta_x, \delta_y)$, and $b$ the righthand side vector.

If the singularity has higher order, the system (6) is singular. In this case we determine which higher order partials also vanish. For each vanishing higher order partial derivative $h$ of $f$, the matrix $A$ is augmented by the row $(h_x(P_i), h_y(P_i))$ and the vector $b$ by the entry $h(P_i)$. This process continues until $D = A^T A$ has full rank. For exceptional values it is possible that $D$ has rank 1 or zero even though no additional partials of $f$ vanish. This means that we happen to approach the singular point crossing a specific algebraic curve given by the symbolic determinant of $D$. In this case, a random perturbation of the point should correct the problem. So far, we have not encountered this problem in practice.

**5.3. Passing to the Transformed Curve.** By a translation of $f$, the coordinate system is centered at the singular point $q$ just found. This may introduce spurious terms that are controlled based on the information obtained during the iteration locating $q$. Recall that the vector $b$ in the iteration contains all partial derivatives that vanish. Consequently, if the partial $h$ appears in $b$, then the corresponding monomial term must be absent in the translated curve. For example, let $\tilde{f}$ be the translation of $f$, and assume that $b$ contains the vanishing partials $f_x$, $f_y$, $f_{xx}$, $f_{yy}$, $f_{xy}$, $f_{xyy}$, and $f_{yyy}$. Then $\tilde{f}$ must not contain the terms $x$, $y$, $x^2$, $y^2$, $xy$, $xy^2$, and $y^3$. Should such terms appear in $\tilde{f}$ with small coefficients, due to numerical imprecision, they are now removed.

Having centered the singularity at the origin, we apply the quadratic transformation (5). Since this transformation maps the line $x = 0$ to infinity, the branch of $\tilde{f}$ we traverse must not have the $y$-axis as tangent. If it does, we rotate $\tilde{f}$ by

$$x' = y \qquad y' = -x$$

before applying the quadratic transformation.

In practice, one applies instead the quadratic transformation

$$x_1 = \frac{x}{y} \qquad y_1 = y$$

in which the $x$-axis becomes the exceptional line. Which quadratic transformation is used is decided based on the current tangent direction of $f$ in the traversal.

**5.4. Branch Orientation.** We give a standard orientation to the curve $f(x, y) = 0$ by the tangent vector $(-f_y, f_x)$. This orientation is not intrinsic in the sense that $-f(x, y) = 0$ is the same curve but with opposite standard direction. Given a consistent traversal direction of a branch, we observe that the standard direction of the curve may reverse at certain singularities. For example, the orientation of $f = y^2 - x^2 - x^3 = 0$ is as shown in Figure 5.3. Consequently, when traversing the curve from $p$ to $q$, we first move in the standard direction, but after the singularity we move in the opposite direction. Figure 5.4 shows that this reversal does not happen at all singularities.

Geometrically, the apparent orientation reversal is understood when considering $f(x, y) = 0$ as the intersection curve of $f(x, y) - z = 0$ and $z = 0$. Along the intersection curve, the projection onto the $x$-$y$ plane of the surface gradient $(f_x, f_y, -1)$ is just the curve normal. Thus the normal reversal that causes the changed standard orientation is merely a rotation of the surface normal in 3-space. Whether a branch suffers this reversal depends on the global topology of the singularity. Briefly, the orientation reverses if the branch intersects an odd number of other branches, with the proper definition of intersection. Nevertheless, a local correspondence between the orientation of $f$ and its proper transform $g$ can be established outside the singularity from which we can deduce whether the standard orientation has reversed, without having to analyze the topology of the singularity.

Let $p = (a_0, b_0)$ be a nonsingular point of $f$, where $a_0 \neq 0$. To $p$ corresponds the point $p_1 = (a_0, b_0/a_0)$ of the transformed curve $g$. Centered at $p$, the curve $f$ has the place

$$\begin{aligned} x(s) &= a_0 + a_1 s + a_2 s^2 + \cdots \\ y(s) &= b_0 + b_1 s + b_2 s^2 + \cdots \end{aligned}$$

and centered at $p_1$, the curve $g$ has the place

$$\begin{aligned} x_1(s) = x(s) &= a_0 + a_1 s + a_2 s^2 + \cdots \\ y_1(s) &= c_0 + c_1 s + c_2 s^2 + \cdots \end{aligned}$$

We assume that the traversal of $f$ at $p$ proceeds by increasing value of $s$. Note that the traversal direction need not agree with the standard orientation $(-f_y, f_x)$. Since

$x(s) = x_1(s)$, the curve and its transform are oriented the same way, and traversing $g$ by increasing $s$ is equivalent to traversing $f$ by increasing $s$.

Since $y_1(s) = y(s)/x(s)$, we divide the two power series and compare the resulting coefficients with the $c_k$. We obtain

$$(7) \qquad \begin{aligned} c_0 &= b_0/a_0 \\ c_1 &= \frac{b_1 a_0 - a_1 b_0}{a_0^2} \\ &\vdots \end{aligned}$$

Since $f$ is not singular at $p$, $g$ is not singular at $p_1$. Hence both curves have a Taylor series at these points so that $a_1$ is proportional to $-f_y$ and to $-g_y$, while $b_1$ is proportional to $f_x$, and $c_1$ is proportional to $g_x$. We now obtain from (7) that

$$\begin{aligned} g_y &= \alpha f_y \\ g_z &= \alpha \frac{x f_x + y f_y}{x^2} \end{aligned}$$

So, if we relate the traversal direction of $f$ to the standard orientation $(-f_y, f_x)$, then $\alpha$ relates the corresponding traversal direction of $g$ to the standard orientation $(-g_y, g_x)$ of $g$, and vice versa. Since the fully desingularized branch cannot experience an orientation reversal, we have a method to maintain consistent traversal direction through singularities.

**5.5. Implementation.** The planar curve traversal has been implemented in Lisp on a Symbolics Lisp machine. A prototype was previously implemented by C. Bajaj on a VAX 8600. Figure 5.4 shows an example of a traversal requiring iterated desingularization, as well as the traversals along the proper transforms in the vicinity of the singularity. For simplicity, the singularity was already positioned at the origin.

## 6. Singularities on Space Curves.

Every algebraic space curve is the birational image of an algebraic plane curve. It follows that the singularities of space curves are not qualitatively different from those of plane curves. Two possibilities exist for tracing through space curve singularities:

1. Construct a birational map from the given space curve to a planar curve, trace the planar curve, and lift the resulting points.

2. Desingularize the space curve directly.

Since the intersection curve in general has degree equal to the product of the surface degrees, the birationally equivalent plane curve must have high degree and may be computationally less tractable. Working with the space curve directly is therefore more attractive. However, desingularizing the space curve directly must address the fact that the curve is given as the intersection of two surfaces. If we are to work with this representation, then we need to apply quadratic transformations that desingularize the curve and the intersecting surfaces as well. Hence the approach to space curve desingularization found in the standard literature, e.g. [5], will not work, and more research is required to work out techniques suitable for computation. We restrict our discussion therefore to the method of reducing the space curve to a plane curve.

The simplest way to map the space curve to a planar curve is by projection. Orthographic projection along a principal axis is done by elimination of a variable, using resultants. Other directions require a rotation of the coordinate system prior to projection. For space curve singularities where at least one of the surfaces has a nonzero gradient, orthographic projection onto the tangent plane is conceptually ideal. The major computational problem would be the inefficiency of the resultant computation for surfaces of high degree. Moreover, branches intersecting the surface normal above or below the tangent plane are also projected and unnecessarily increase the complexity of the singularity.

A different method to map a space curve to the plane is to find a rational surface containing the curve, parameterizing this surface, and then substituting the parametric equations into one of the implicit surface equations, say $g$. This method has the advantage of treating all singularities, not only those at a nonzero surface gradient. We describe the method below and give several examples. It has not been implemented yet.

**6.1. Monoid and Cone Representation.** It is well known that every algebraic space curve $f \cap g$ can be represented as the intersection of a monoid and a cone, [13]. A *monoid* is a rational surface of degree $m$ that contains an $m - 1$ fold point. Simple examples include all planes, quadrics, and the Steiner surface. When the $m - 1$ fold point is brought to the origin, the monoid equation takes the form

$$w H_{m-1}(x, y, z) + H_m(x, y, z) = 0$$

where $H_{m-1}$ is homogeneous of degree $m - 1$ and $H_m$ is homogeneous of degree $m$.

We will be interested in determining the monoid containing a given space curve $f \cap g$ and its parametric representation. We will not determine the cone, since it is not needed. Moreover, the parameterization of the monoid is incompatible with the cone equation. The procedure for determining the monoid is based on the projective form of the surfaces $f$ and $g$. As in [12], we proceed as follows.

First, homogenize $f(x, y, z)$ and $g(x, y, z)$ so as to obtain $F(w, x, y, z)$ and $G(w, x, y, z)$. As long as $w \neq 0$, the curve $F \cap G$ is identical to $f \cap g$. We select one of the base points of the projective coordinate system, say $(1, 0, 0, 0)$, as the $m - 1$ fold monoid point. This implies using $w$ as the main variable in the computation below. The base point $(0, 1, 0, 0)$ would correspond to selecting $x$ as the main variable, and so on.

We write both $F$ and $G$ as polynomials in the main variable, $w$,

$$
\begin{aligned}
F &= u_n w^n + u_{n-1} w^{n-1} + \cdots + u_1 w + u_0 \\
G &= v_{n'} w^{n'} + v_{n'-1} w^{n'-1} + \cdots + v_1 w + v_0
\end{aligned}
$$

Without loss of generality we assume that $n \geq n' > 1$. We compute the polynomials

$$F_1 = u_n w^{n-n'} G - v_{n'} F$$
$$G_1 = (u_0 G - v_0 F)/w$$

Note that both $F_1$ and $G_1$ contain the intersection curve of $F$ and $G$.

Both $F_1$ and $G_1$ have degree at most $n - 1$ in $w$. If one of them is linear in $w$, then we stop; we have found the monoid equation. If neither is linear, then we repeat the calculation using $F_1$ and $G_1$ in place of $F$ and $G$. Since at each step the maximum degree in $w$ is lowered at least by one, the computation derives the monoid equation after at most $n$ steps in the form

$$w H_{m-1}(x, y, z) + H_m(x, y, z) = 0$$

The monoid is parameterized by intersecting it with lines through the $m - 1$ fold point. Let $a : b : c$ be the direction ratios of these lines, then the monoid is parameterized by

$$w(a, b, c) = -H_{m-1}(a, b, c)/H_m(a, b, c)$$
$$x(a, b, c) = a$$
$$y(a, b, c) = b$$
$$z(a, b, c) = c$$

This parameterization is *projective*, that is, $(a, b, c)$ are the coordinates of a two-dimensional projective parameter space.

The parametric forms are now substituted into the equation of $G$ and give the desired plane curve, in homogeneous form.

**6.2. Examples.** We illustrate the method with two examples. First, consider the intersection curve of the cylinder $F = x^2 + z^2 + 2zw = 0$ and the sphere $G = x^2 + y^2 + z^2 + 4zw = 0$. The intersection curve is an irreducible degree 4 space curve with a nodal singularity at the origin shown in Figure 6.1.

The cylinder is a monoid with the $m - 1$ fold point at the origin $(1, 0, 0, 0)$. Since the point of interest on the space curve is the origin, we determine a different monoid whose $m - 1$ fold point is not the origin. We choose $(0, 0, 0, 1)$, making $z$ the main variable. Accordingly, we compute

$$F_1 = G - F = y^2 + 2zw$$

The parameterization of $F_1$ is then

$$z = -2a/c^2$$
$$w = a$$
$$x = b$$
$$y = c$$

Substitution into $G$ yields the plane curve

$$b^4 + 4a^2(c^2 - b^2) = 0$$

Dehomogenizing with $a = 1$ yields $b^4 - 4(c^2 - b^2) = 0$. This curve is shown in Figure 6.2.

As a more complicated example, consider the intersection of the torus $(x^2 + y^2 + z^2 - w^2)^2 + 8w^2(z^2 - x^2 - y^2 - w^2) + 16w^4 = 0$ with the ellipsoid $36(z - w)^2 + 4(y - w)^2 + 9x^2 - 36w^2 = 0$. The monoid computation, as described, yields a surface of degree 12 in 4 steps. Its equation contains an extraneous factor of degree 4. Substitution into the ellipsoid equation thus yields a plane curve of degree 16 that factors into a degree 2 component, a degree 6 component, and a degree 8 component.

A better solution to the problem is to parameterize the ellipsoid since it is a monoid. To do so, we first translate the coordinate system to the point $(1, 0, 1, 2)$, which is on the ellipsoid, and parameterize with $w$ as main variable. This results in a plane curve of degree 12 that could not be factored by Macsyma.

**6.3. Remarks.** Only in the plane curve case do we know of a simple algebraic procedure achieving complete desingularization at minimal computational cost. The projection of space curves to planar curves seems to require significant machinery, for determining the monoid equation, and for eliminating extraneous components introduced in the process. In the case of monoid/cone intersection these extraneous components are all lines, hence would be simple to exclude. The cone is determined by a resultant computation, and yields a homogeneous polynomial in three variables. Unfortunately, its equation is unsuitable for substituting the monoid parameterization. In simple cases, there exist certain reparameterizations that circumvent this problem.

If one of the surfaces $f$ or $g$ is known to be rational it can be advantageous to parameterize it directly. In the case of quadrics this amounts to a coordinate translation that brings one of the base points of the coordinate system onto the surface. In the case of cubics, [2] gives parameterization algorithms. Many other surfaces, including the torus, are also rational with known standard parameterizations. Direct parameterization side-steps a potentially lengthy monoid derivation. It should be noted, however, that the resulting plane curve is not necessarily of minimum degree.

It appears that there are simple quadratic transformations of space that achieve surface/surface intersection desingularization much as in the planar case. More research is needed to explore the exact effect of these transformations, and the issues involved in realizing them by computation.

**Acknowledgements.** Much of this research was done jointly with C. Bajaj, J. Hopcroft, and R. Lynch.

## REFERENCES

[1] S. ABHYANKAR, *Desingularization of Plane Curves*, Proc. Symposia in Pure Math. 40 (1983) Part 1, 1–45.

[2] S. ABHYANKAR AND C. BAJAJ, *Automatic Rational Parameterization of Curves and Surfaces II: Cubics and Cubicoids*, Comp. Aided Design, to appear.

[3] C. BAJAJ, C. HOFFMANN AND J. HOPCROFT, *Tracing Algebraic Curves: Plane Curves*, Tech. Rep. CSD-TR-637 (1987), Purdue University.

[4] G. COLLINS, *Quantifier elimination for real closed fields by cylindrical algebraic decomposition*, 2nd GI Conf. on Aut. Thy. and Formal Lang. (1975), Springer Lect. Notes in Comp. Sci. 33, 134–183.

[5] A. EMCH, *Reduction of Singularities of Space Curves and Surfaces*, in *Algebraic Geometry*, V. Snyder et al., eds., Chelsea Publications 1970, 252–256.

[6] R. FAROUKI, *The Characterisation of Parametric Surface Sections*, Comp. Vision, Graphics and Image Proc. 33 (1986) 209–236.

[7] R. FAROUKI AND V. RAJAN, *On the Numerical Condition of Bernstein Polynomials*, IBM Res. Rept. RC 12626 (1987).

[8] C. HOFFMANN AND R. LYNCH, *Following Space Curves Numerically*, Tech. Rept. CSD-TR-684 (1987), Purdue University.

[9] M. PRATT AND A. GEISOW, *Surface/surface intersection problems*, in *The Mathematics of Surfaces*, J. A. Gregory, ed., Clarendon Press, Oxford 1986, 117–142.

[10] J. SEMPLE AND G. KNEEBONE *Algebraic Curves*, Clarendon Press, Oxford 1959.

[11] J. SCHWARTZ AND M. SHARIR, *On the Piano Movers' Problem II*, in *Planning, Geometry, and Complexity of Robot Motion*, J. Schwartz, M. Sharir, J. Hopcroft, eds., Ablex Publishing, Norwood, N.J., 1987, 51–96.

[12] V. SHARIR AND C. SISAM, *Analytic Geometry of Space*, H. Holt and Co., New York 1914.

[13] B. L. VAN DER WAERDEN, *Einführung in die algebraische Geometrie*, 2nd edition, Springer Verlag 1973.

[14] R. WALKER *Algebraic Curves*, Springer Verlag 1950.

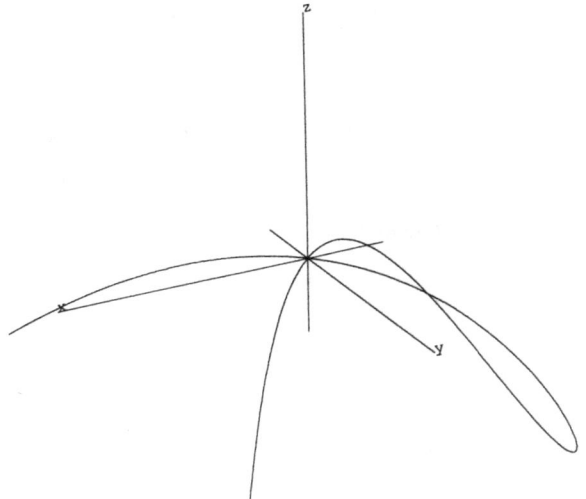

Figure 3.1

$z + y^2 - x^3 \ \cap \ z + x^2$
Normal Crossing

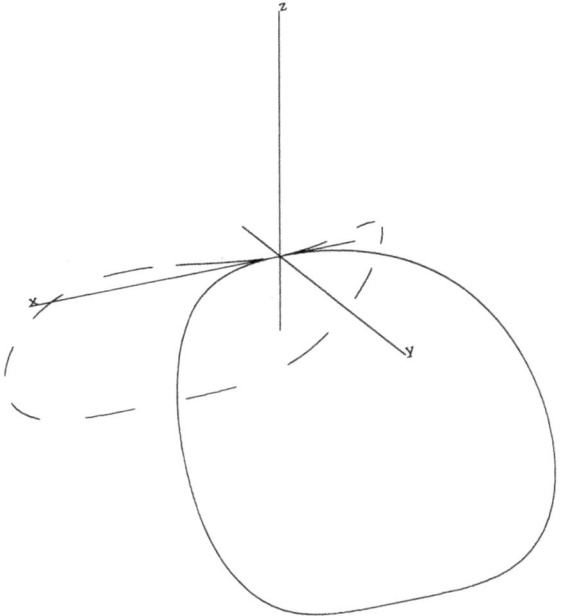

Figure 3.2

$z + x^4 + y^4 \ \cap \ z + y^2$
Tacnode

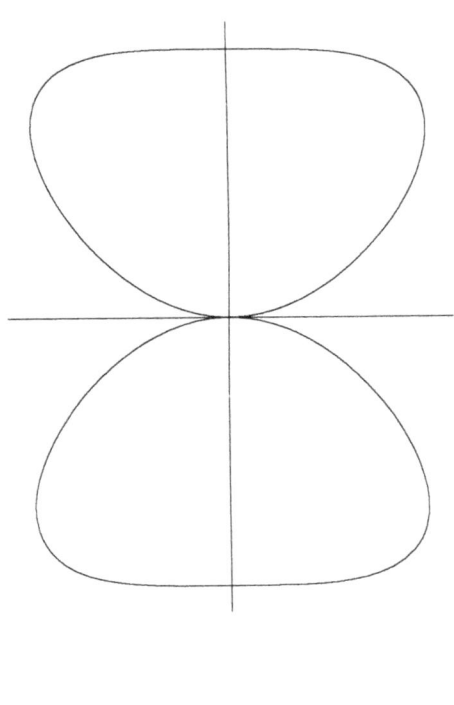

Figure 3.3

$y^2 - x^4 - y^4$
Touching Branches

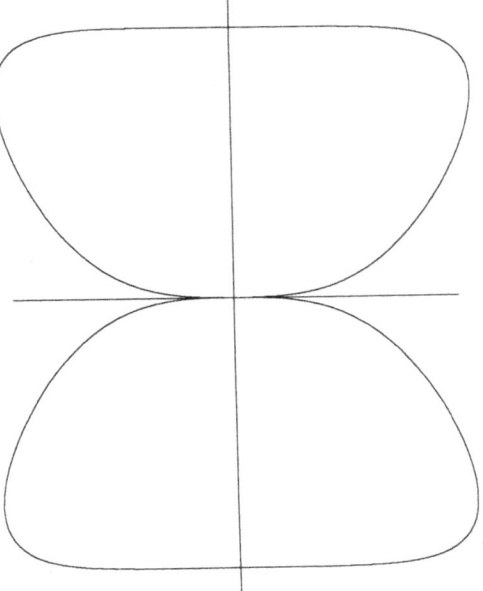

Figure 3.4

$y^2 - x^6 - y^6$
Intersecting Branches

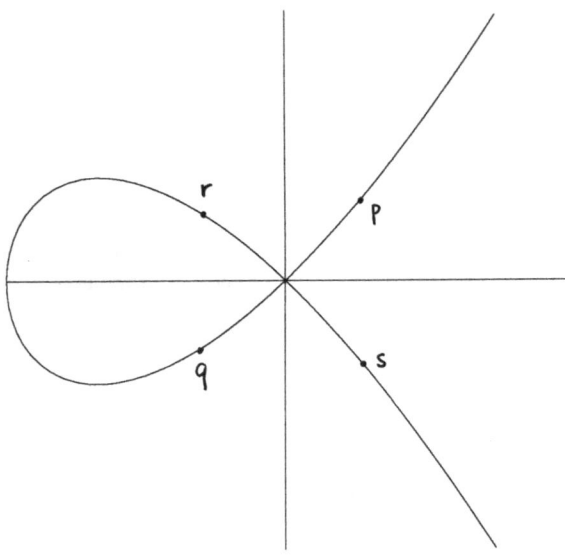

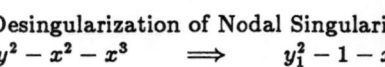

Figure 5.1

Desingularization of Nodal Singulari⬛

$$y^2 - x^2 - x^3 \implies y_1^2 - 1 - x$$

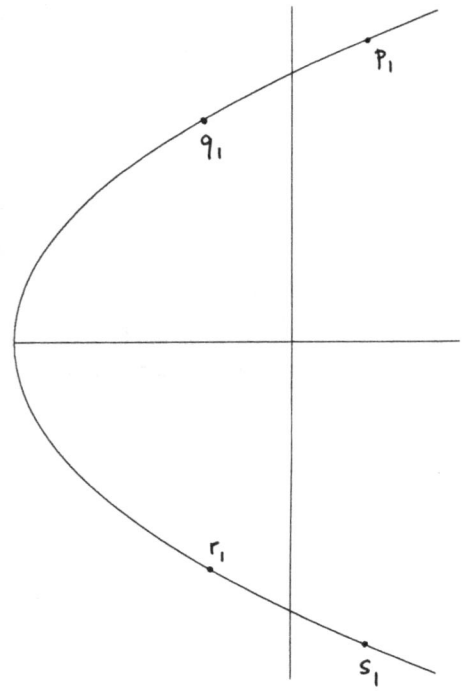

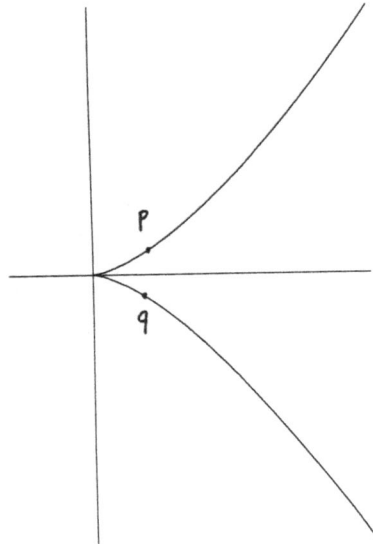

Figure 5.2

Desingularization of Cuspidal Singularity

$$y^2 - x^3 \implies y_1^2 - x_1$$

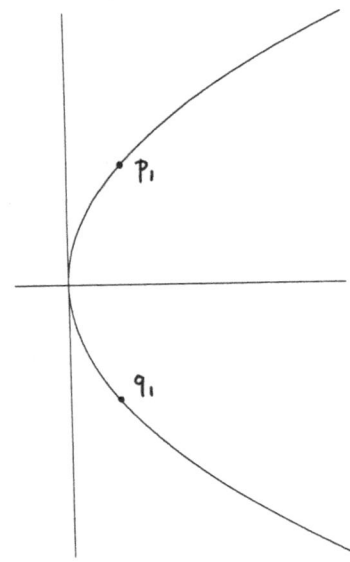

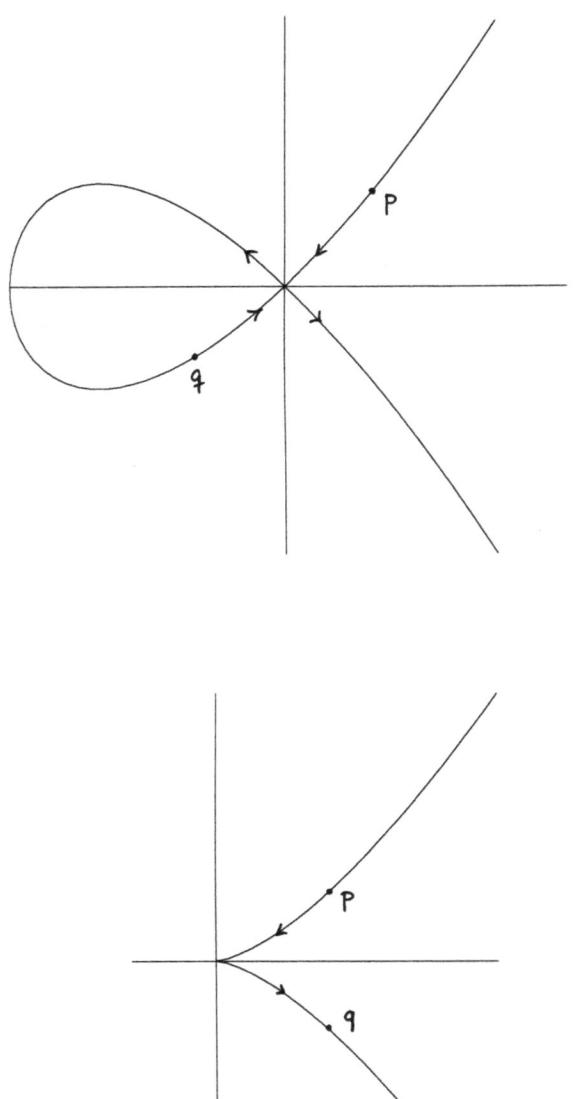

Figure 5.3

$$y^2 - x^2 - x^3$$
Orientation Reversal at Singularity

Figure 5.4

$$y^2 - x^3$$
No Orientation Reversal at Singularity

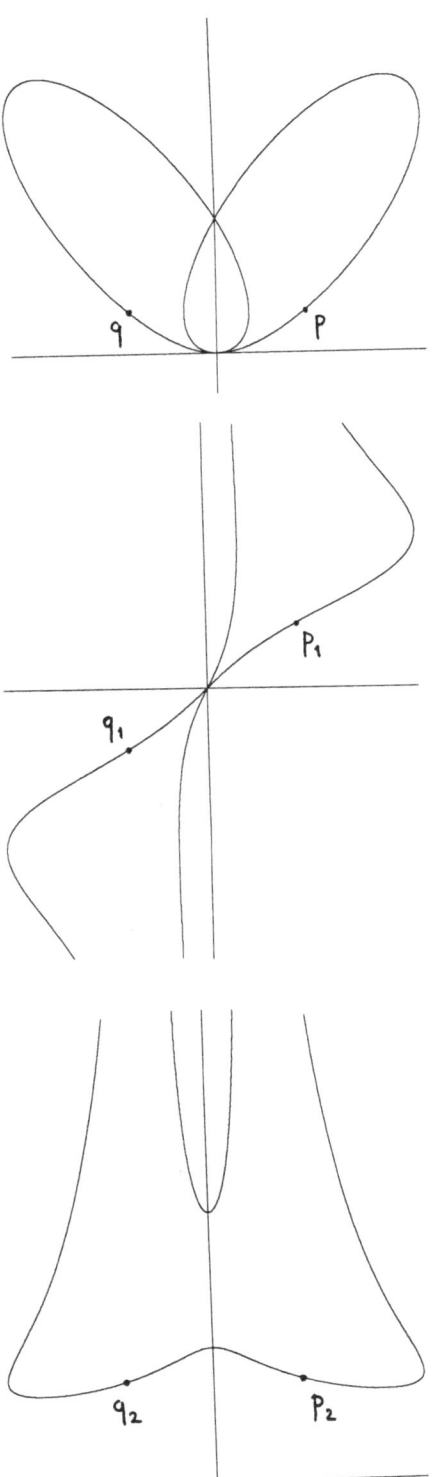

Figure 5.5

Recursive Desingularization
$$2x^4 - 3x^2y + y^2 - 2y^3 + y^4$$
$$\Downarrow$$
$$2x_1^2 - 3y_1x_1 + y_1^2 - 2y_1^3x_1 + y_1^4x_1^2$$
$$\Downarrow$$
$$2 - 3y_2 + y_2^2 - 2y_2^3x_2^2 + y_2^4x_2^4$$

122

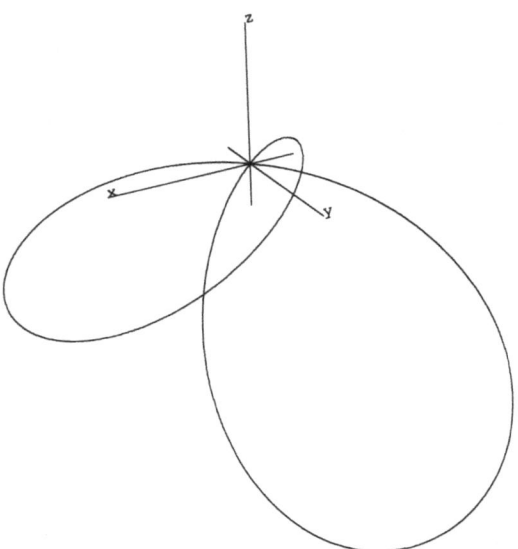

Figure 6.1

Cylinder - Sphere Intersection
$$x^2 + y^2 + z^2 + 4z \cap x^2 + z^2 + 2z$$

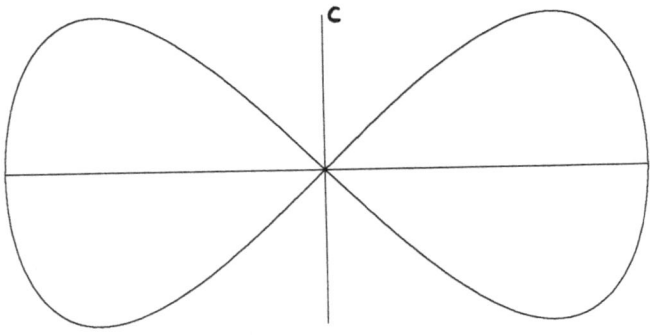

Figure 6.2

Corresponding Plane Cur
$$b^4 - 4(c^2 - b^2)$$

# PERFORMANCE OF SCIENTIFIC SOFTWARE

E.N. HOUSTIS, J.R. RICE, C.C. CHRISTARA AND E.A. VAVALIS*
Department of Computer Sciences
Purdue University

## Abstract

We review performance methodologies used for the evaluation of scientific software in von Neumann architectures. A prototype evaluation facility for second order elliptic partial differential equation (PDE) solvers is described which realizes the main objectives of these methodologies. Finally, the results of an evaluation study for a new class of spline collocation solvers for elliptic PDEs are presented.

## 1. Introduction

In this paper, we address the issue of performance evaluation of mathematical software. Specifically, we are interested in comparing and evaluating different items of mathematical software which perform the same task and require input of similar nature. Unfortunately, the underlying mathematical problem is often unsolvable and various experimental approaches have been formulated for its solution. They are usually based on subjective criteria and parameter domains, which are often subject to dispute. Nevertheless, it has been recognized that performance evaluation is the only way to bring some rational order to the world of scientific software. In Section 2, we justify the need for evaluation studies and review the mathematical formulation of the evaluation problem. In Section 3, we list the design principles and objectives which an experimental evaluation process must satisfy. In Section 4, we describe a prototype evaluation facility for second order elliptic partial differential equation (PDE) solvers. Finally, in Section 5, we present the results of an evaluation study for a new class of elliptic PDE solvers obtained using this facility. We consider only traditional von Neumann machines, performance evaluation in parallel computing environment is much more complex [8] although the same principles apply.

* This work was supported in part by Air Force Office of Scientific Research grant 84–0385 and by the Strategic Defense Initiative, ARO grant DAAL03–86–K–0106.

## 2. The Performance Evaluation Problem

The problem of selecting an efficient algorithm arises frequently in different situations. For example, the numerical treatment of many scientific applications involves the solution of basic mathematical problems. This creates the need to select appropriate mathematical software for their solution. In developing a software library, there is the need of identifying the most efficient algorithms for each task and implementing them. In both situations, the results of a scientifically designed evaluation can be of great assistance. At present, several research and development efforts exist for the development of expert systems for scientific computation. Performance evaluation results should be incorporated as part of the knowledge base of such systems. Finally, performance evaluation studies can assist the developers of new algorithms to compare their performances against existing ones without repeating the experiment for all of them. This process can also result in the tuning of existing implementations or in a better overall understanding of them.

### 2.1 *Performance Evaluation Model*

An abstract model for the performance evaluation of software [14] involves five parameters, the *problem space* or *population of problems* (**P**), an *algorithm space* or *collection of software* (**A**) that can be used to solve each problem $x$ in **P**, a *space of performance measures* (**M**) used to define the performance of each algorithm $A \in$ **A**, *performance norms* used to estimate algorithm performance from various measurements, and a *selection mapping* (**S**($x$)), that determines the *effective, good* or *best* algorithm for given problems. Figure 2.1 due to Rice [14] summarizes an abstract framework for the evaluation of software. It is easy to realize that the evaluation problem is often unsolvable. The objective of this paper is to implement scientifically sound methodologies for the experimental evaluation of elliptic PDE solvers.

An experimental evaluation process depends very much on the choice of the five parameters involved. Although their selection is subjective, standard scientific guidelines can be followed to design such experiments. For example, in the design of a population of problems, we must assure that it represents reality, it includes subclasses of problems with well known features and problems which depend continuously on a number of continuous parameters [11].

**P:**   Problem space or a population of problems.

**A:**   Algorithm space or a collection of software that can be used to solve $x \in$ **P**.

**M:**   Performance measure space.

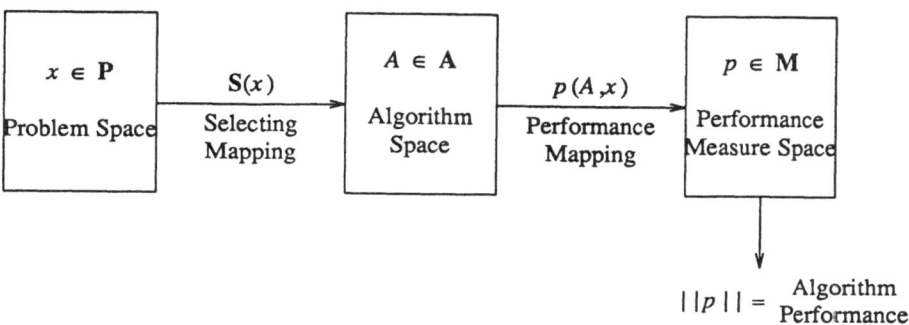

**Figure 2.1.**   An abstract framework of performance evaluation of software.

The latter characteristic allows the computation of performance profiles for each algorithm over a range of parameter values that correspond to specific problem features. Many algorithms require input with specific properties (e.g., smoothness of functions) to perform optimally or at all. We believe that the population of problems should contain elements that test such algorithmic assumptions. Finally, it should be large enough to allow statistically meaningful results. Using this model, one can address several questions related to the performance of algorithms.

The results of such studies can be used to identify *an algorithm out of a subclass* $A_0$ *which maximizes performance over a problem subset* $\mathbf{P}_0$. Another question of practical significance is *the determination of subclasses of problems for which an algorithm* $A \in$ A *is effective , poor or unreliable*. Evaluation studies can be used to *verify hypotheses*. For example, Houstis and Rice [10] have applied an experimental approach to verify that high order methods are more effective than low order methods for the numerical solution of linear, elliptic PDEs on rectangles and which have singularities or near singularities. Another usage of evaluation studies is the classification and ranking of various algorithms (programs) that perform the same task. For example, Rice's study [16] led to the following ranking of linear algebraic equation solvers for Galerkin's equations (see [16] for precise definitions of the terminology):

1. Jacobi CG, Jacobi CG + RCM, SOR + RCM, SOR
2. Linpack, Yale Sparse + ND9
3. Yale Envelope
4. Yale Sparse + MD
5. Yale Envelope + RCM, Yale Sparse
6. Sparse GE
7. Yale Sparse + RCM
8. Yale Envelope + MD

## 3. Design Principles of Experimental Evaluation

As in any scientific experiment, the evaluation of mathematical software must adhere to some basic principles. In [4] Crowder et al, list and describe such principles. They state that experimental evaluation studies must

(a) *explicitly state assumptions, hypotheses, objective criteria and measurements,*

(b) *their results must be reproducible,*

(c) *others must be able to access easily the experimental data,*

and

(d) *complete descriptions of the experimental apparatus must be given.*

Finally, they state that such experiments must allow the developers of new software to perform the experiments using their software without repeating the experiment for the existing items. The ELLPACK group at Purdue has developed an evaluation facility for elliptic PDE solvers that adheres to the above principles, it is described next.

## 4. Facility For the Performance Evaluation of Elliptic PDE Solvers

We now describe an evaluation facility for second order elliptic PDE solvers, which realizes the main objectives of the methodologies presented above. The facility consists of a population of elliptic PDEs [17], [18], [19]; a collection of algorithms known with the acronym ELLPACK [1], [18]; a subsystem for defining computational experiments [1]; the ELLPACK system that carries out each experiment and measures various performance indicators [18]; a subsystem that collects and organizes the performance data [3];

and, finally, a subsystem that performs a statistical test and generates performance curves [3], [18]. In the rest of this section we present an overview of this system.

## 4.1 Population of Elliptic PDEs

We are interested in the performance of algorithms for solving the following type of elliptic PDE problems:

$$Lu \equiv Au_{xx} + Bu_{xy} + Cu_{yy} + Du_x + Eu_y + Fu = f$$

defined on $\Omega \subseteq R^2$ and subject to boundary conditions $Bu \equiv \alpha u_x + \beta u_x = g$ defined on the boundary of $\Omega$.

For this purpose, a population of elliptic PDEs was formed [18], [19]. It is divided into two groups, based on the geometry of the domain of definition. There are 56 PDE problems defined on rectangular regions [5]. Most of them depend on one or more parameters that control various features of the problem. By selecting appropriate parameter values, 189 PDE problem instances were selected as a "basic" population [5]. They exist both in machine and human readable form. The population covers problems with various features. The latter are classified with respect to *Operator Type* (Poisson, Helmholtz, Self-Adjoint, Constant coefficients, General), *Boundary Conditions* (Dirichlet, Neumann, Mixed), and *Solution Features* (Entire, Analytic, Singular, Peak, Oscillatory, Boundary Layer, Wave Front, Singularities, Irregular, Discontinuities, Computationally Complex). Figure 4.1 depicts a human readable form of the population, together with contour plots of the solution. Also the features of the problem and/or solution are specified. Figure 4.2 depicts a machine readable form of the population. The problem features have been encoded, the basic problems are defined as records, while problems involving parameters are called macros.

## 4.2 ELLPACK Elliptic Solvers

A collection of over 50 software modules for PDE problems is available known with the acronym ELLPACK. The modules are classified according to the task they perform. An elliptic PDE solver in the ELLPACK environment can be viewed either as *multi-segment*, that is, consisting of three different modules that perform the three basic tasks of discretizing the PDE problem, formatting (or indexing) the corresponding algebraic equations and solving them or as a *single-segment*.

128

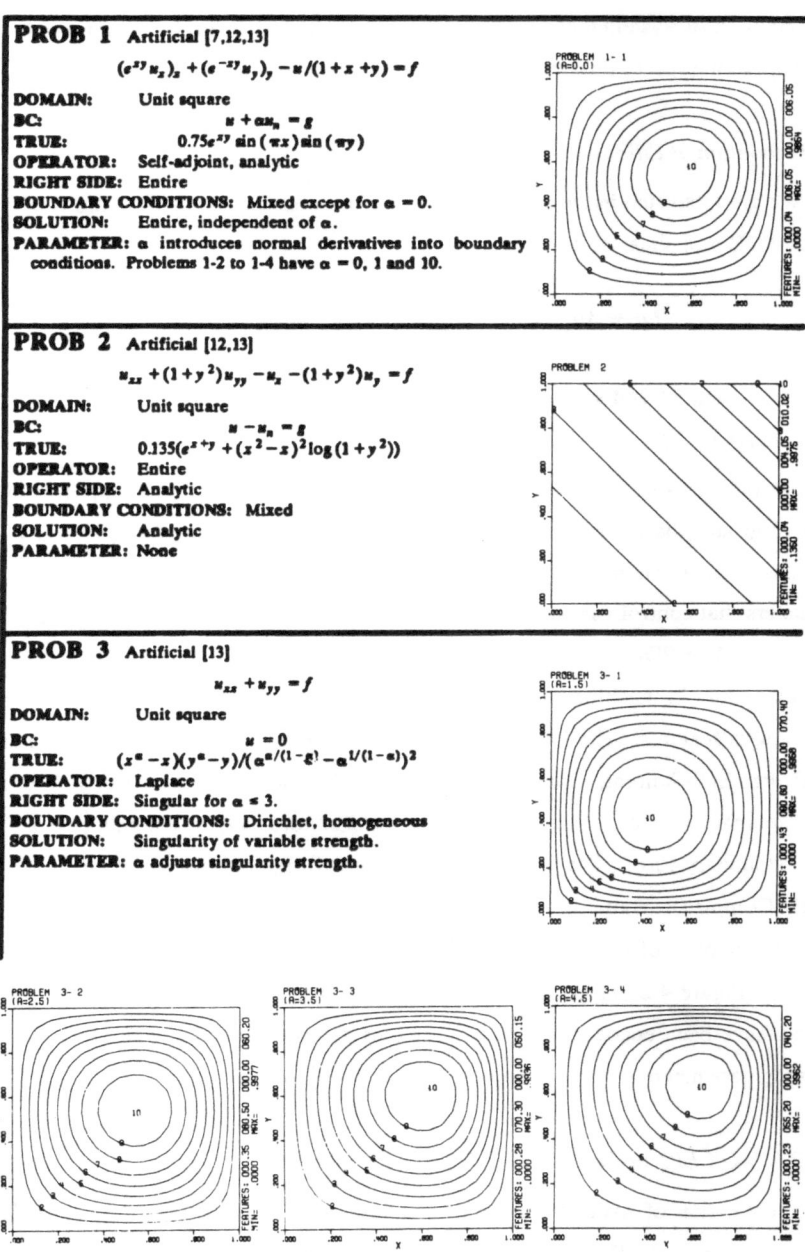

**PROB 1** Artificial [7,12,13]

$$(e^{xy}u_x)_x + (e^{-xy}u_y)_y - u/(1+x+y) = f$$

**DOMAIN:**    Unit square
**BC:**    $u + \alpha u_n = g$
**TRUE:**    $0.75e^{xy}\sin(\pi x)\sin(\pi y)$
**OPERATOR:**    Self-adjoint, analytic
**RIGHT SIDE:** Entire
**BOUNDARY CONDITIONS:** Mixed except for $\alpha = 0$.
**SOLUTION:**    Entire, independent of $\alpha$.
**PARAMETER:** $\alpha$ introduces normal derivatives into boundary conditions. Problems 1-2 to 1-4 have $\alpha = 0$, 1 and 10.

**PROB 2** Artificial [12,13]

$$u_{xx} + (1+y^2)u_{yy} - u_x - (1+y^2)u_y = f$$

**DOMAIN:**    Unit square
**BC:**    $u - u_n = g$
**TRUE:**    $0.135(e^{x+y} + (x^2-x)^2\log(1+y^2))$
**OPERATOR:**    Entire
**RIGHT SIDE:** Analytic
**BOUNDARY CONDITIONS:** Mixed
**SOLUTION:**    Analytic
**PARAMETER:** None

**PROB 3** Artificial [13]

$$u_{xx} + u_{yy} = f$$

**DOMAIN:**    Unit square
**BC:**    $u = 0$
**TRUE:**    $(x^\alpha - x)(y^\alpha - y)/(\alpha^{\alpha/(1-\alpha)} - \alpha^{1/(1-\alpha)})^2$
**OPERATOR:**    Laplace
**RIGHT SIDE:** Singular for $\alpha \leq 3$.
**BOUNDARY CONDITIONS:** Dirichlet, homogeneous
**SOLUTION:**    Singularity of variable strength.
**PARAMETER:** $\alpha$ adjusts singularity strength.

**Figure 4.1.** Human readable form of the population [19].

```
**************
*  problem 2    *
**************
*
*          000.04    000.00    004.05    010.02
*          2000200000020
1          uxx + (1.+y*y)uyy − ux − (1.+y*y) uy = f(x,y)
2          u + ux = 0.27*exp(y)                    on x=0.
2          u − ux = 0.                             on x=1.
2          u + uy = 0.27*exp(x)                    on y=0.
2          u − uy = 0.135*(alog(2.)−1.)*(x*x−x)**2  on y=1.
3     function true(x,y)
3     true = 0.135*(exp(x+y)+(x*x−x)**2*alog(1.+y*y))
3     return
3     end
3     function f(x,y)
3     f = 0.135*( (−4.*x*x*x+18.*x*x−14.*x+2.)*alog(1.+y*y)
3   $      − 2.*((x*x−x)**2)*(y*y+y**3+y−1.)/(1.+y*y) )
3     return
3     end
-----------------------------------------------------------------
**************
*  macro   4    *
**************
*          2020021000020
1          uxx + uyy = 6.*x*y*exp(x+y)*(x*y+x+y−3.)
2          u − (&a)*(y−y*y)ux = g(y)               on x=0.
2          u=0.                                    on x=1.
2          u=0.                                    on y=0.
2          u=0.                                    on y=1.
3     function true(x,y)
3     true = 3.*exp(x+y)*x*y*(1.−x)*(1.−y)
3     return
3     end
3     function g(y)
3     temp = (y−y*y)**2
3     g = −(&a)*temp*3.*exp(y)
3     return
3     end
-----------------------------------------------------------------
```

**Figure 4.2.** Machine readable form of the population [18].

```
**************
*  problem 4    *
**************
* parameter set     1(a=0.0)
*               000.02    004.00    000.00    006.00
*               2020021002002
1               uxx + uyy = 6.*x*y*exp(x+y)*(x*y+x+y-3.)
2               u=0. on  x=0.
2               u=0. on  x=1.
2               u=0. on  y=0.
2               u=0. on  y=1.
3       function true(x,y)
3       true = 3.*exp(x+y)*x*y*(1.-x)*(1.-y)
3       return
3       end
-
* parameter set     2(a=0.1)
*               000.02    004.00    000.00    006.00
expand 4/0.1/
-
* parameter set     3(a=1.0)
*               000.02    004.00    000.00    006.00
expand 4/1.0/
-
* parameter set     4(a=10.)
*               000.02    004.00    000.00    006.00
expand 4/10./
-
* parameter set     5(a=1000.)
*               000.02    004.00    000.00    006.00
expand 4/1000./
----------------------------------------------------------------
```

**Figure 4.2**. (Continued)

Notice that each solver can be specified in encoded form by specifying the appropriate module number, such as

/Discretization Mod./Indexing Mod./Solution Mod./

for a multi-segment solver and

/Discretization Mod.///

for a single-segment solver. The solver /42, degree = 2, nderv = 1/14/20 corresponds to Galerkin method based on quadratic splines with natural indexing and LINPACK band solver, while /9/// corresponds to an FFT solver.

The ELLPACK has a very high level user interface [18] that allows the user to specify the problem and method in a natural form. Figure 4.3 depicts an ELLPACK program which specifies a PDE problem, two multi-segment solvers and one single-segment solver. Note that the ELLPACK preprocessor generates the appropriate Fortran program needed, which is hundreds of lines of code.

### *Discretization modules*

A number of well known discretization modules are implemented in ELLPACK, both finite difference and finite element schemes. This is the current list of discretizations modules:

| | |
|---|---|
| 1. Collocation (General Domain) | 6. Spline Galerkin |
| 2. Hermite Collocation | 7. 5 Point Star |
| 3. Hodie | 8. 5 Point Star (General Domain) |
| 4. Hodie Helmholtz | 9. 7 Point Star 3D |
| 5. Interior Collocation | 10. Spline Collocation: One Step |

The indicated numbering is used for the encoded specification of the method.

### *Indexing modules*

There are modules that order the discrete equations according to well known schemes that various linear algebraic equation solvers expect. Ordering routines from ITPACK, YALEPACK and LINPACK software packages are included. This is the current list of indexing modules:

| | |
|---|---|
| 20. As Is | 24. Nested Dissection |
| 21. Hermite Collorder | 25. Red Black |
| 22. Interior Collorder | 26. Reverse Cuthill McKee |
| 23. Minimum Degree | |

Again the numbering is used for the encoded specification of this module.

### *Solution modules*

Banded direct, sparse direct and iterative methods are used available for solving the discretized PDE equations. Many of these solvers have been obtained from LINPACK, YALEPACK and ITPACK software packages. The following is the current list of solution modules:

| | |
|---|---|
| 30. Band GE | 40. Symmetric SOR CG |
| 31. Band GE No Pivoting | 41. Reduced System SI |
| 32. Envelope LDLT | 42. Reduced System CG |
| 33. Envelope LDU | 43. Sparse GE No Pivoting |
| 34. Linpack Band | 44. Sparse LDLT |
| 35. Linpack SPD Band | 45. Sparse LU Compressed |
| 36. SOR | 46. Sparse LU Pivoting |
| 37. Jacobi SI | 47. Sparse LU Pivoting |
| 38. Jacobi CG | |
| 39. Symmetric SOR SI | |

*Triple modules*

Finally a number of very fast algorithms are included as single segment modules which are called triples in the jargon of ELLPACK. The domain of applicability of these methods is usually restricted. ELLPACK currently includes FFT, Hodie, Multigrid, Marching methods and some general deferred correction spline collocation methods.

| | |
|---|---|
| 63. Dyakanov CG | 70. Marching Algorithm |
| 64. Dyakanov CG 4 | 71. Multigrid MG00 |
| 65. FFT 9 Point | 72. P2C0 Triangles |
| 66. Fishpak Helmholtz | 73. Set |
| 67. Hodie FFT | 74. Set U by Bicubics |
| 68. Hodie FFT 3D | 75. Set U by Blending |
| 69. Hodie 27 Point 3D | 76. Spline Collocation: Two Step (General, Interior) |

*Utility procedures*

A number of procedures exist for performing related tasks to the solution process of the discrete PDE models. The following is the current list of such procedures.

| | |
|---|---|
| 80. Display Matrix Pattern | 87. Set Unknowns For Hodie Helmholtz |
| 81. Domain Fill | 88. Set Unknowns For 5 Point Star |
| 82. Eigenvalues | 100. Domain Processor |
| 83. Plot Collocation Points | 101. Interpolation |
| 84. Remove | 102. Plot |
| 85. Remove Bicubic BC | 103. Table |
| 86. Remove Blended BC | 104. Max |

```
*
*
            Sample Ellpack Program
*
*

opti.
            max x points = 33      $ max y points = 33
            time                   $ memory
equa.
            uxx + uxy + uyy + fux(x, y)*ux + uy + u = f(x, y)
bound.
            u        = true(x, y)                  on x = 0
            ux       = xtrue(x, y)                 on x = 1
            uy       = ytrue(x, y)                 on y = 0
            u + uy   = true(x, y) + ytrue(x, y)    on y = 1

for.
            do 1 n = 1, 5
                   gridps = 2**n + 1
grid.              gridps x points & gridps y points
disc.              spline collocation
inde.              as is
solu.              envelope ldu
out.               max(error)

inde.              minimum degree
solu.              spars lu pivoting
out.               table(error)

triple.            twostep spline collocation
out.               max(error)
for.
    1              continue
subprograms.
*FORTRAN subprograms for f , fux , true , xtrue , ytrue functions.
end.
```

**Figure 4.3**.  An ELLPACK program.


**4.3** *Performance Measure Space*

To access the performance of each method, a number of performance indicators must be computed. Table 4.1 lists such indicators appropriate for von Neumann machines and PDE problems.

| Parameter | Description |
|-----------|-------------|
| nx | Number of $x$ grid points |
| ny | Number of $y$ grid points |
| h | max $(1/(nx - 1), 1/(ny - 1))$ |
| nunk | number of unknowns |
| rerr(20 × 20) | relative maximum error at 20 × 20 grid |
| rerrmax | relative maximum error at the nodes |
| errl$_2$ | $L_2$-error |
| resmax | residual maximum error at 20 × 20 |
| resmxr | residual maximum error |
| reslr | $L_2$-residual error |
| solmax | Solution maximum out |
| nit | Number of iterations |
| mem | Memory requirements |
| $t_1$ | Discretization time |
| $t_2$ | Indexing time |
| $t_3$ | Solution time |
| tt | Total time $(t_1 + t_2 + t_3)$ |

**Table 4.1.** Performance measure space for von Neumann machines.

**4.4** *A Performance Evaluation System*

Figure 4.4 shows a block view of the experimental evaluation facility based on ELLPACK. The first subsystem reads a file with the problem identification numbers and the file of the encoded methods, together with the mesh specifications. Table 4.2 presents the information needed for an experiment 18 PDE problems and 10 methods. Only 144 different PDE problems are defined as some methods do not apply to some problems. These illegal combinations are automatically skipped by the performance evaluation system.

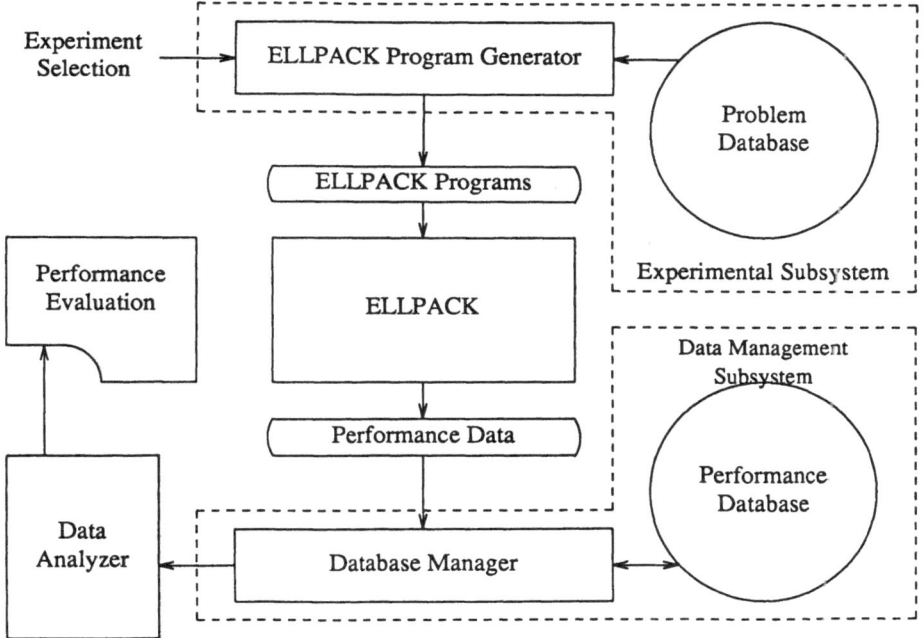

**Figure 4.4.** The performance evaluation system based on ELLPACK.

| Population of PDE problems | Methods |
|---|---|
| 1–1  8–2  22–1 | 56/14/63 |
| 3–1  9–1  33–1 | 5,5 |
| 4–1  10–2  41–3 | 9,9 |
| 5–1  10–3  47–2 | 17,17 |
| 5–4  11–2  50–1 | 33,33 |
| 6–1  17–2  54–2 | 57/// |
| | 5,5 |
| | 9,9 |
| | 17,17 |
| | 33,33 |

**Table 4.2.** The specifications of a PDE performance evaluation experiment with 144 different runs.

| Encoded Experiments | |
|---|---|
| 901–1//9,9/58,homo=.true./14/20//$/ $ | 7–1//19,19/58,homo=.true./14/20//$/1 $ |
| 5–1//9,9/58,homo=.true./14/20//$/ $ | 10–2//19,19/58,homo=.true./14/20//$1 $ |
| 5–4//9,9/58,homo=.true./14/20//$/ $ | 10–3//19,19/58,home=.true./14/20//$/ $ |
| 6–1//9,9/58,homo=.true./14/20//$/ $ | 41–1//19,19/58,homo=.true./14/20//$/ $ |
| 7–1//9,9/58,homo=.true./14/20//$/ $ | 901–1//29,29/58,home=.true./14/20//$/ $ |
| 10–2//9,9/58,homo=.true./14/20//$/ $ | 5–1//29,29/58,homo=.true./14/20//$/ $ |
| 10–3//9,9/58,homo=.true./14/20//$/ $ | 5–4//29,29/58,homo=.true./14/20//$/ $ |
| 41–1//9,9/58,homo=.true./14/20//$/ $ | 6–1//29,29/58,home=.true./14/20//$/ $ |
| 901–1//19,19/58,homo=.true./14/20//$/ $ | 7–1//29,29/58,homo=.true./14/20//$/ $ |
| 5–1//19,19/58,homo=.true./14/20//$/ $ | 10–2//29,2958,homo=.true./14/20//$/ $ |
| 5–4//19,19/58,homo=.true./14/20//$/ $ | 10–3//29,29/58,homo=.true./14/20//$/ $ |
| 6–1//19,19/58,homo=.true./14/20/$/ $ | 41–1//29,2958,homo=.true./14/20//$/ $ |

**Figure 4.5.** Input to the ELLPACK program generator unit for the performance evaluation experiment of Table 4.2.

Figure 4.5 shows the encoded file with this information. Notice that this input specifies 144 different Fortran programs to be executed. Each program produces the 17 performance data items of Table 4.1 that are collected for each problem/method pair. This is accomplished by the *Data Management Subsystem*. Figure 4.6 depicts the organization and display of these data as done by the *data manager*. The *data analyzer* is capable of producing performance curves for various methods and the same problem by applying a linear least squares fit to the corresponding data. Figure 4.7 shows an example of performance curves obtained automatically by the data analyzer.

10 2 b4p2unix/56/14/63/                            (a=50.0, b=0.5)

| nx | ny | h | nunk | rerr | rerrmax | err12 | resmax | resmxr |
|----|----|------|------|---------|---------|---------|---------|---------|
| 5  | 5  | 0.25e+00 | 36 | 0.17e+01 | 0.17e+01 | 0.24e+00 | 0.17e+01 | 0.10e+01 |
| 9  | 9  | 0.13e+00 | 100 | 0.16e+00 | 0.12e+00 | 0.17e−01 | 0.51e+01 | 0.10e+01 |
| 17 | 17 | 0.63e−01 | 324 | 0.87e−02 | 0.51e−02 | 0.96e−03 | 0.11e+02 | 0.10e+01 |
| 33 | 33 | 0.31e−01 | 1156 | 0.49e−03 | 0.14e−03 | 0.11e−03 | 0.13e+02 | 0.10e+01 |

| res12 | solmax | nit | mem | tt | t1 | t2 | t3 |
|--------|---------|-----|-------|--------|-------|------|-------|
| 0.51e+01 | 0.86e−01 | 0 | 10028 | 1.72 | 0.86 | 0.07 | 0.7 |
| 0.51e+02 | 0.54e−01 | 0 | 27636 | 6.42 | 2.21 | 0.07 | 4.1 |
| 0.29e+02 | 0.62e−01 | 0 | 89252 | 39.04 | 7.28 | 0.07 | 31.6 |
| 0.55e+02 | 0.63e−01 | 0 | 318084 | 371.26 | 26.04 | 0.12 | 345.1 |

10 2 b4p2unix/56/36,itmax=500/                      (a=50.0, b=0.5)

| nx | ny | h | nunk | rerr | rerrmax | err12 | resmax | resmxr |
|----|----|------|------|---------|---------|---------|---------|---------|
| 9  | 9  | 0.13e+00 | 100 | 0.16e+00 | 0.12e+00 | 0.17e−01 | 0.51e+01 | 0.10e+01 |
| 19 | 19 | 0.56e−01 | 400 | 0.53e−02 | 0.17e−02 | 0.66e−03 | 0.11e+02 | 0.10e+01 |
| 29 | 29 | 0.36e−01 | 900 | 0.85e−03 | 0.82e−03 | 0.16e−03 | 0.13e+02 | 0.10e+01 |

| res12 | solmax | nit | mem | tt | t1 | t2 | t3 |
|--------|---------|-----|-------|--------|-------|------|-------|
| 0.15e+02 | 0.54e−01 | 21 | 6636 | 18.17 | 2.17 | 0.17 | 15.8 |
| 0.33e+02 | 0.62e−01 | 19 | 26156 | 68.67 | 8.67 | 0.17 | 59.8 |
| 0.49e+02 | 0.62e−01 | 22 | 58676 | 172.17 | 19.83 | 0. | 152.3 |

**Figure 4.6.** Display of the performance data for two problem/method pairs. Each uses problem 10−2 with parameters (50, 0.5) and was run on the Berkeley 4.2 UNIX compiler (F77).

Problem 41- 3

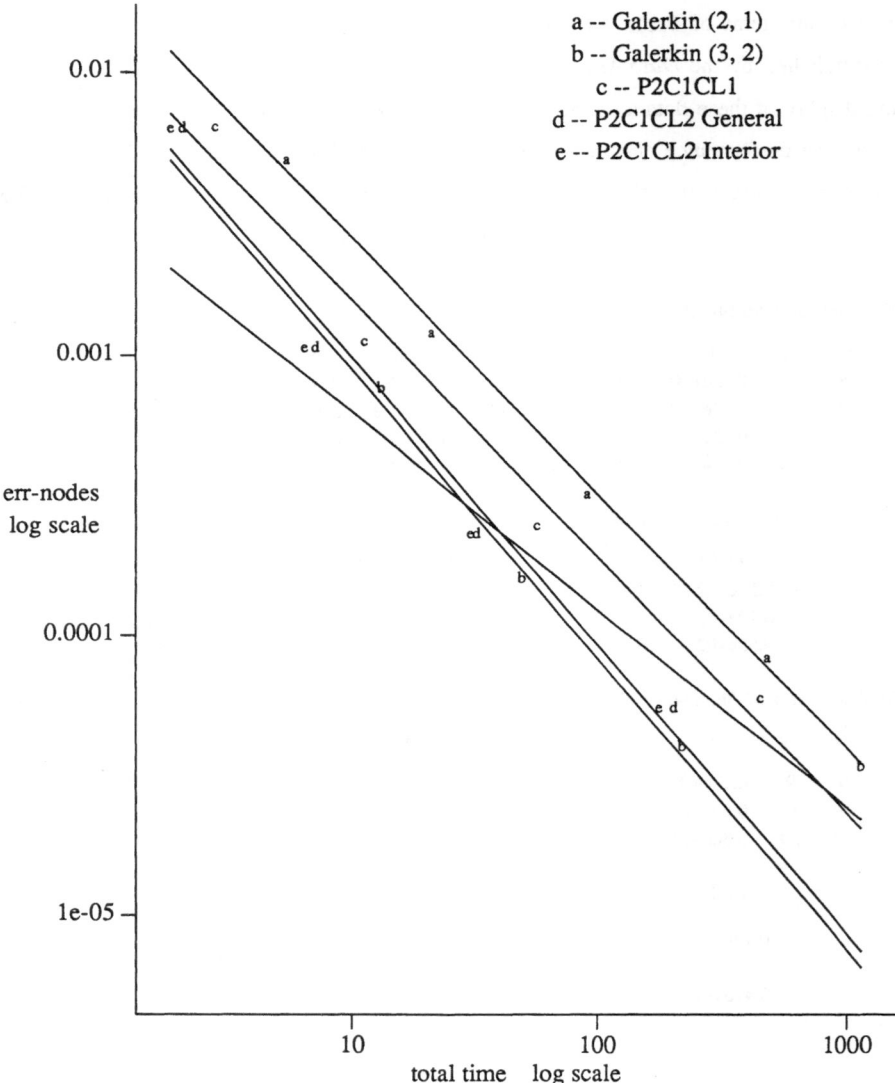

**Figure 4.7.** Performance curves of method/problem pairs obtained by linear least squares fitting of the raw data.

In order to rank these methods over a problem or problems, a statistical test is applied to the slopes of specified performance curves [6]. For example, the slopes are measured for each method and PDE. These slopes are ranked from 1 (the best) to the number of methods tested and the ranks are averaged over the number of problems considered.

| plot: x=nx, y=time3 | | | | | | |
|---|---|---|---|---|---|---|
| problems= 1–1  4–1  5–1  5–4  6–1  7–1  10–2  10–3  41–1 | | | | | | |
| | | | slope   time 3   -nx | | | |
| Methods | ave rank | minimum | 1st quart | median | 3rd quart | maximum |
| b4p2unix/58,homo=.true./14/36/ | 1.00 | 2.4e+00 | 2.5e+00 | 2.5e+00 | 2.6e+00 | 2.8e+00 |
| b4p2unix/58,homo=.true./14/35/ | 2.00 | 2.7e+00 | 2.8e+00 | 2.8e+00 | 2.8e+00 | 3.0e+00 |
| b4p2unix/58,homo=.true./14/33/ | 3.00 | 2.8e+00 | 2.9e+00 | 2.9e+00 | 3.0e+00 | 3.0e+00 |
| b4p2unix/58,homo=.true./16/54/ | 4.56 | 3.2e+00 | 3.2e+00 | 3.4e+00 | 3.4e+00 | 3.5e+00 |
| b4p2unix/58,homo=.true./16/53/ | 5.00 | 3.2e+00 | 3.3e+00 | 3.4e+00 | 3.4e+00 | 3.8e+00 |
| b4p2unix/58,homo=.true./16/60/ | 6.22 | 3.4e+00 | 3.5e+00 | 3.5e+00 | 3.5e+00 | 3.6e+00 |
| b4p2unix/58,homo=.true./14/53/ | 7.22 | 3.3e+00 | 3.5e+00 | 3.6e+00 | 3.7e+00 | 3.9e+00 |
| b4p2unix/58,homo=.true./14/54/ | 8.22 | 3.6e+00 | 3.6e+00 | 3.6e+00 | 3.7e+00 | 3.8e+00 |
| b4p2unix/58,homo=.true./14/63/ | 9.67 | 3.5e+00 | 3.5e+00 | 3.8e+00 | 3.9e+00 | 4.2e+00 |
| b4p2unix/58,homo=.true./14/20/ | 9.67 | 3.7e+00 | 3.7e+00 | 3.7e+00 | 3.8e+00 | 3.9e+00 |
| b4p2unix/58,homo=.true./14/45/ | 9.89 | 3.6e+00 | 3.7e+00 | 3.7e+00 | 3.9e+00 | 3.9e+00 |
| b4p2unix/58,homo=.true./14/60/ | 11.56 | 3.8e+00 | 3.9e+00 | 3.9e+00 | 3.9e+00 | 4.1e+00 |

Confidence level =

ave rank differences significant with 80.00, 90.00, 95.00, 97.50, 99.00 percent certainty if greater than 4.68, 5.15, 5.55, 5.92, 6.36.

**Figure 4.8.**    Ranking of methods based on the slopes of *total execution time* vs. *nx* performance curves. Nine problems (identified at the top) have been solved with twelve PDE solvers (identified on the left).

This test also produces a confidence interval for the ranking. If the difference in the ranking of two methods A and B corresponds to confidence level L, then we say that the confidence level for method A outperforming method B is L. Figure 4.8 shows an example of the ranking output generated based on *total execution time* (time 3) vs *nx*. Figure 4.9 indicates the order of convergence of various methods for several PDE problems produced automatically by the data analyzer subsystem.

| 10 2 b4p2unix/56/14/63 | | | (a=50.0, b=0.5) |
|---|---|---|---|
| nx | nx1 | cerr | cerrmax |
| 5 | 9 | 4.02 | 4.51 |
| 9 | 17 | 4.58 | 4.97 |
| 17 | 33 | 4.34 | 5.42 |

| 10 2 b4p2unix/56/14/63 | | | (a=50.0, b=0.5) |
|---|---|---|---|
| nx | nx1 | cerr | cerrmax |
| 9 | 19 | 4.56 | 5.70 |
| 19 | 29 | 4.33 | 1.72 |

| 10 2 b4p2unix/57,home=.true./// | | | (a=50.0, b=0.5) |
|---|---|---|---|
| nx | nx1 | cerr | cerrmax |
| 5 | 9 | 3.92 | 4.51 |
| 9 | 17 | 4.50 | 4.97 |
| 17 | 33 | 4.25 | 6.26 |

**Figure 4.9.** Order of convergence of various ELLPACK methods. The convergence orders based on the relative error and maximum relative error are denoted by cerr and cerrmax.

## 5. Case Study: Performance Evaluation of Spline Collocation Methods For Elliptic PDEs.

In this section we present the results of an evaluation study for a new class of elliptic collocation solvers using the ELLPACK facility. The *objectives* of this study is to *assess the performance of various linear solvers for this new class of collocation discretizations.*

### 5.1 Description of Algorithm Space

For completeness, we describe very briefly the general idea of this new class of methods, [7], [9], [12]. Assuming the general boundary value problem $Lu = f$, $Bu = g$ on $\Omega \equiv [a, b] \times [c, d]$, we approximate $u$ by a tensor product of smooth splines $(S_n \equiv \mathbf{P}_{n,\Delta} \cap C^{n-1}(\Omega))$. Throughout we denote by $I$, an appropriate interpolant of $u$ in $S_n$. Then the following high order asymptotic relations at appropriate *collocation points* are derived for $I$ :

$$LI = \tilde{f}(I) + O(h^n) \quad \text{or} \quad L'I = f + O(h^n)$$

where $h = \max(|\Delta_x|, |\Delta_y|)$, with $|\Delta|$ denoting the maximum mesh length. Based on these asymptotic relations, a collocation approximation $A \in S_n$ is determined by forcing it to satisfy either

(one step)   $[L'A - f]_P = 0$ including appropriate auxiliary boundary conditions

or

(two step)   $[LA^{(1)} - f]_P = 0$ and $[LA^{(2)} - \tilde{f}(A^{(1)})]_P = 0$ including appropriate auxiliary/boundary conditions

where $P$ are the collocation points. It turns out that the approximation error of $A$ or $A^{(2)}$ behaves asymptotically as $O(h^n)$, which is optimal with respect to interpolation using smooth splines. Currently, we have implemented and analyzed such methods for $n = 2, 3, 5$. These methods are referred to by the acronym PXCYCLS where $X$ indicates the degree of splines, $Y$ their smoothness and $S$ the formulation (one or two step). Special efficient implementations of these methods exist for PDEs with homogeneous boundary conditions and Helmholtz operators. They are referred as *Interior* and *Helmholtz* modules respectively. Table 5.1 indicates the new ELLPACK collocation modules, using this method.

| MODULES | | APPLICABILITY | |
|---------|---|-----------|---|
| | | Operator | Boundary Conditions |
| Interior | P2C1CL2 | General | Homogeneous D/N* |
| General | P2C1CL1 | General | Mixed |
| General | P2C1CL2 | General | Mixed |
| Helmholtz | P3C2CL1 | Helmholtz | D/N |
| Interior | P3C2CL1 | General | Homogeneous D/N |
| Interior | P3C2CL2 | General | Homogeneous D/N |
| General (1-D) | P5C4CL1 | General | Mixed |

*D = Dirichlet type, N = Neumann type

**Table 5.1.** New ELLPACK collocation modules based on smooth splines approximations.

**5.2** *Performance evaluation study I: The experiment*

In this study, we consider the solution of algebraic equations obtained by quadratic spline collocation discretization method applied to linear elliptic partial differential equations. These equations are solved by the following ELLPACK modules:

SOR:        A program for SOR iteration.

SSOR SI:       A program for SOR accelerated by semi-iteration.

JACOBI CG:      A Jacobi method accelerated by CG.

LINPACK BAND:    A LU factorization technique with partial pivoting for banded matrices.

BAND GE NO PIVOTING: Modified version of LINPACK BAND.

ENVELOPE LDU:    A LDU factorization program for matrices in envelope form.

SPARSE GE NO PIVOTING: Computes a LU factorization of a matrix using a fast storage conserving non-symmetric scheme.

SPARSE LU UNCOMPRESSED: Computes LU factorization of a matrix using a fast non-symmetric storage scheme.

SPARSE LU PIVOTING:  A sparse Gauss elimination program with column pivoting.

All the above modules apply to quadratic equations in the natural ordering. One more ordering in ELLPACK was used.

MD (Minimum Degree): Minimizes the number of nonzero elements introduced during Gauss elimination.

This experimental study was conducted on the collection of problem from the ELLPACK population. The actual problems used are listed in the Appendix. The quadratic spline collocation equations are generated by ELLPACK programs P2C1CL1 and P2C1CL2 (interior = satisfies exactly B.C). Figures 5.1 and 5.2 depict the structures of these systems of algebraic equations.

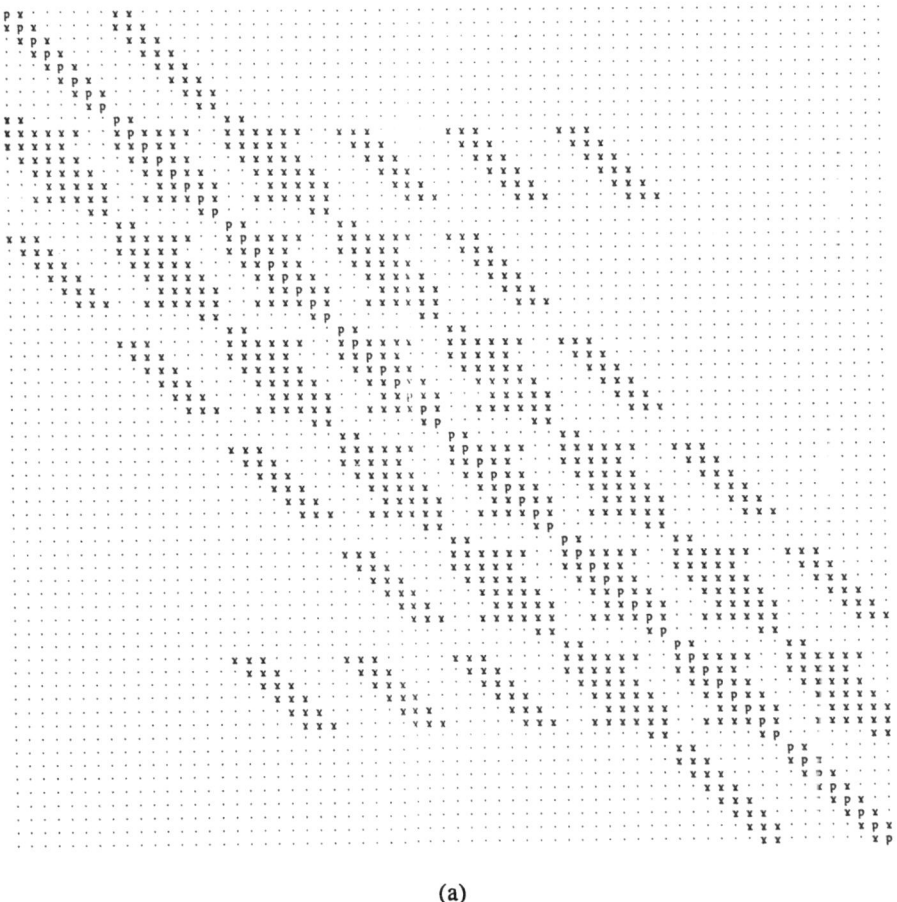

(a)

**Figure 5.1.** Pattern of nonzero elements in the quadratic collocation method equations (P2C1CL1) for (a) the natural ordering and (b) the minimum degree ordering. grid. The problem is $Lu \equiv u_{xx} + u_{yy} + u_{yy} + u_x + u_y + u$ with a 7×7 grid.

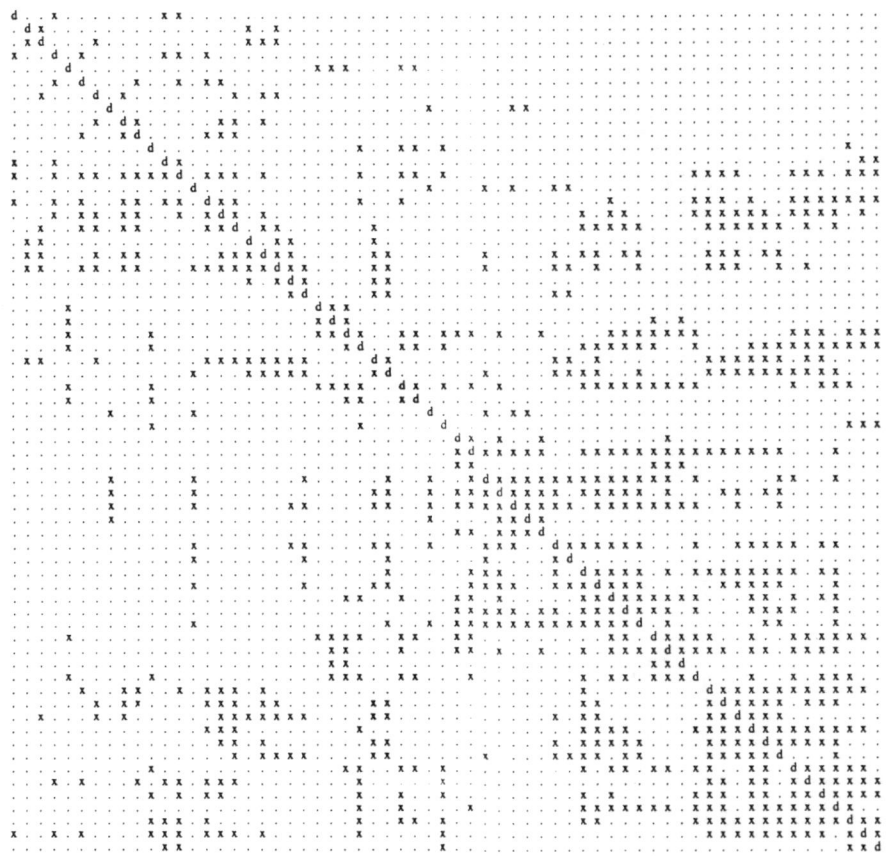

(b)

**Figure 5.1.** (continued)

145

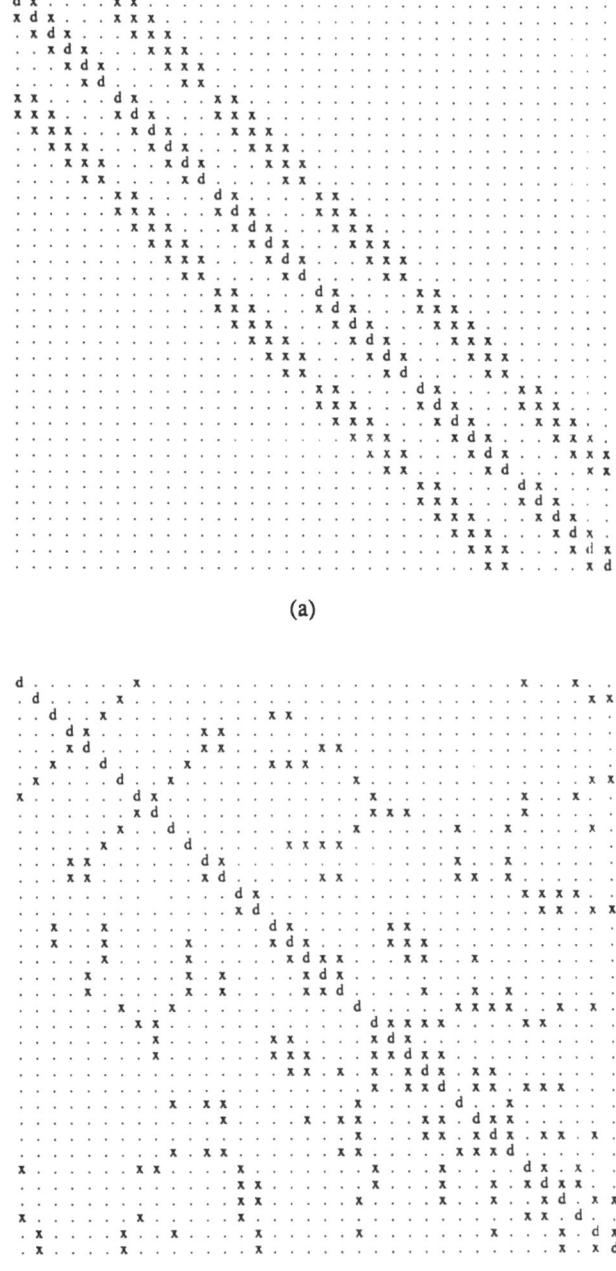

(a)

(b)

**Figure 5.2.**  Pattern of nonzero elements in the quadratic collocation method equations (P2C1CL2) for (a) the natural ordering and (b) the minimum degree ordering. The same problem as in Figure 5.1 was used with a 7×7 grid.

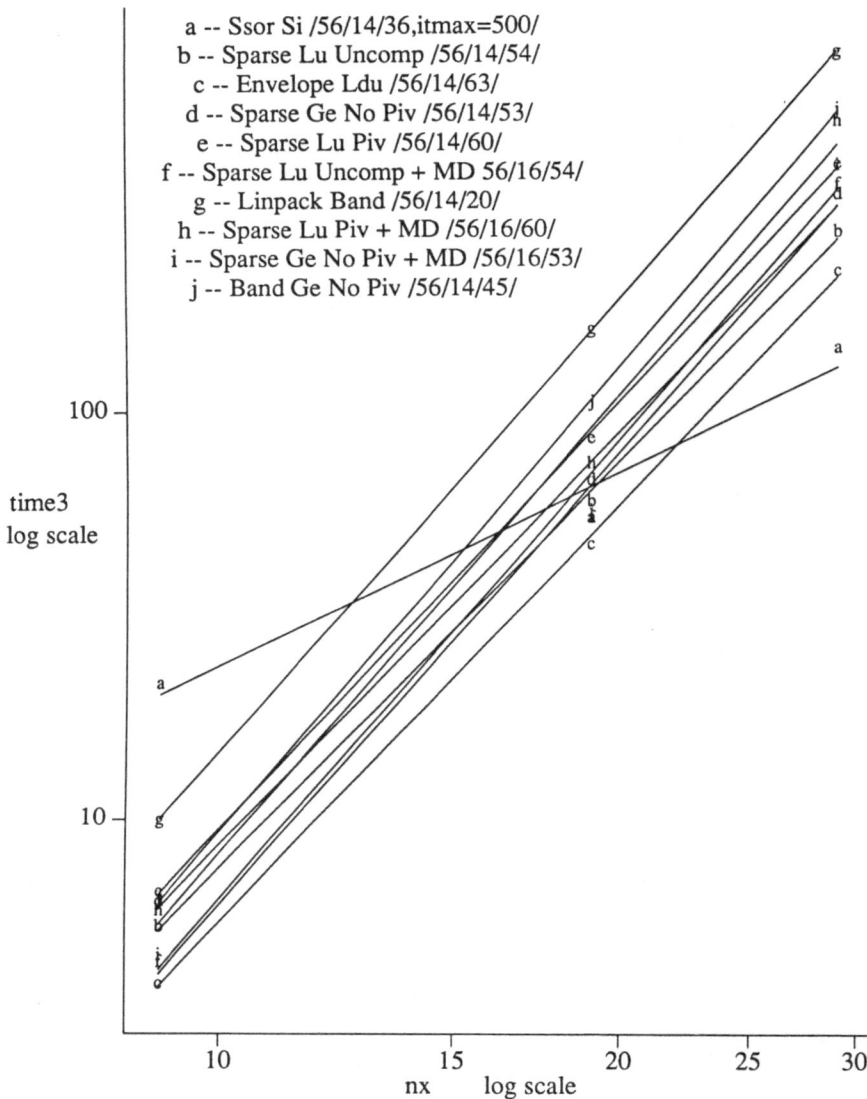

**Figure 5.3.** Performance of ten ELLPACK equation solvers on a log-log scale based on solution time (time 3) in seconds.

## 5.3 *Discussion of the performance data*

The criterion of performance is the computer time (*time 3* in VAX 8600 seconds) required to solve the linear system. All PDEs were solved on a uniform square $N \times N$ grid and the resulting system is of order $(N + 2)^2$ for P2C1CL2 (interior) and P2C1CL1 with half bandwidth $N$ and $5N + 6$ respectively. Figure 5.3 indicates the performance of the ten equation solvers on a log-log scale, based on *time 3*. It is worth noticing that several iterative methods are applicable to quadratic collocation equations, which, for large systems, tend to outperform some of the direct methods. If we consider the *slope* of *time 3* vs $N$ as the primary measure of performance, then the method with the smallest slope is the most efficient asymptotically for large $N$. Based on this slope, the ranking of the 10 modules considered is shown in Table 5.2.

| Method | Average Rank | Median Slope | Sample run times for Prob. 1–1 | | |
|---|---|---|---|---|---|
| | | | 9×9 grid | 19×19 grid | 29×29 grid |
| SSOR SI | 1.00 | 1.9 | 2.1 | 5.6 | 14.3 |
| Sparse LU Uncomp | 2.22 | 3.3 | 0.6 | 6.1 | 27.4 |
| Envelope LDU | 3.56 | 3.4 | 0.4 | 4.8 | 22.0 |
| Sparse GE No PIV | 3.89 | 3.4 | 0.6 | 6.8 | 33.8 |
| Sparse LU PIV | 4.78 | 3.5 | 0.7 | 8.6 | 40.3 |
| Sparse LU Uncomp + MD | 6.67 | 3.7 | 0.5 | 5.5 | 35.7 |
| Linpack Band | 7.56 | 3.7 | 1.0 | 15.9 | 75.0 |
| Sparse LU PIV + MD | 7.67 | 3.7 | 0.6 | 7.5 | 51.0 |
| Sparse GE No PIV + MD | 8.11 | 3.7 | 0.5 | 5.8 | 39.9 |
| Band GE No PIV | 9.56 | 3.8 | 0.6 | 10.6 | 54.3 |

**Table 5.2.** Average rankings and median slopes of 10 modules for solving the quadratic collocation equations (General P2C1CL1). Actual times for one problem are given in the right 3 columns. All timings are in seconds on a VAX 8600 using double precision.

Table 5.2A gives the pairwise confidence levels for rankings in Table 5.2. If the $i, j$ entry in Table 5.2A is $L$, then the confidence level for the hypothesis that method $i$ outperforms method $j$ is greater than $L$.

| | 1 | 2 | 3 | 4 | 5 | 6 | 7 | 8 | 9 | 10 |
|---|---|---|---|---|---|---|---|---|---|---|
| 1. SSOR SI | – | | | | | | | | | |
| 2. Sparse LU Uncomp | <80 | – | | | | | | | | |
| 3. Envelope LDU | <80 | <80 | – | | | | | | | |
| 4. Sparse GE No PIV | <80 | <80 | <80 | – | | | | | | |
| 5. Sparse LU PIV | 80 | <80 | <80 | <80 | – | | | | | |
| 6. Sparse LU Uncomp + MD | 99 | 90 | <80 | <80 | <80 | – | | | | |
| 7. LINPACK Band | 99 | 99 | 80 | <80 | <80 | <80 | – | | | |
| 8. Sparse LU PIV + MD | 99 | 99 | 80 | 80 | <80 | <80 | <80 | – | | |
| 9. Sparse GE No PIV + MD | 99 | 99 | 95 | 90 | <80 | <80 | <80 | <80 | – | |
| 10. Band GE No PIV | 99 | 99 | 99 | 99 | 95 | <80 | <80 | <80 | <80 | – |

**Table 5.2A.** Pairwise confidence levels for the rankings in Table 5.2.

These data on confidence levels break the modules into two groups, the first 5 and the second 5, which are "weakly" ordered (confidence levels less than 80%) internally. While the border between the two groups is a little weak (i.e., module Sparse LU PIV is better than Sparse LU Uncomp + MD with less than 80% confidence level), the extremes are clearly well separated. That is, SSOR SI is better than all the second group with 99% confidence level. Likewise, Band GE No PIV is worse than all the first group with 99% confidence level (only 95% for Sparse LU PIV). Had we made an experiment involving more PDE problems, we believe that we would have strong statistical support for the group rankings given below.

The data supports the following subjective rankings in groups, based on timing on 29×29 grid:

1. SSOR SI, Envelope LDU.

2. Sparse LU Uncompressed, Sparse GE No Pivoting, Sparse LU Uncompressed + MD, Sparse GE No Pivoting + MD.

3. Sparse LU Pivoting + MD, Band GE No Pivoting.

4. LINPACK Band.

The performance of 11 modules was considered based on approximate memory requirements. The algebraic equations were generated by the General P2C1CL1 module. Table 5.3 lists these estimates over three different grids. The data confirms the expected superior performance of the iterative methods vs the direct ones.

| Method | Memory | | |
|--------|--------|--|--|
| | 9×9 grid | 19×19 grid | 29×29 grid |
| Linpack Band | 15500 | 122000 | 409500 |
| Sparse LU PIV | 26762 | 123002 | 312762 |
| Sparse LU PIV + MD | 26174 | 117960 | 312309 |
| Band GE No PIV | 10300 | 81200 | 272700 |
| Sparse LU Uncomp | 9520 | 70000 | 229480 |
| Sparse LU Uncomp + MD | 8344 | 59916 | 228614 |
| Envelope LDU | 21600 | 96400 | 194400 |
| Sparse GE No PIV + MD | 4228 | 20684 | 72224 |
| Sparse GE No PIV | 3317 | 21312 | 66007 |
| Jacobi CG | 2320 | 4000 | 6500 |
| SSOR SI | 600 | 2400 | 5400 |

**Table 5.3.** Approximate memory requirements of 11 modules for solving the quadratic collocation equations (General P2C1CL1).

Similar experiments were carried out for the equations generated by *Interior* P2C1CL2 and *General* P2C1CL2 module. In these cases, the algebraic systems generated are smaller. Table 5.4 shows the ranking of 12 modules for solving the equations obtained from *Interior* P2C1CL2. The pairwise confidence level of the ranking is given in Table 5.4A.

| Method | Average Rank | Medium Slope | Sample run times for Prob. 1–1 | | |
|--------|---------|-------|----------|------------|------------|
| | | | 9×9 grid | 19×19 grid | 29×29 grid |
| SSOR SI | 1.00 | 2.5 | 0.3 | 2.0 | 5.4 |
| Jacobi CG | 2.11 | 2.8 | 0.3 | 2.3 | 9.1 |
| SOR | 2.89 | 2.9 | 0.3 | 2.6 | 7.9 |
| Sparse LU Uncomp + MD | 4.56 | 3.2 | 0.1 | 0.9 | 3.9 |
| Sparse GE No PIV + MD | 5.00 | 3.3 | 0.1 | 0.9 | 3.9 |
| Sparse LU PIV + MD | 6.22 | 3.5 | 0.1 | 0.9 | 4.3 |
| Sparse GE No PIV | 7.22 | 3.5 | 0.1 | 1.3 | 6.2 |
| Sparse LU Uncomp | 8.33 | 3.6 | 0.1 | 1.2 | 5.8 |
| Envelope LDU | 9.56 | 3.5 | 0.0 | 0.9 | 4.3 |
| LINPACK Band | 9.67 | 3.7 | 0.1 | 1.0 | 5.3 |
| Band GE No PIV | 9.89 | 3.7 | 0.1 | 1.0 | 4.9 |
| Sparse LU PIV | 11.56 | 3.9 | 0.1 | 1.5 | 7.8 |

**Table 5.4.** Average rankings and median slopes of 12 modules for solving the quadratic collocation equations (Interior P2C1CL2). Actual times for one problem are given in the right 3 columns. All timings are in seconds on a VAX 8600 using double precision.

|  | 1 | 2 | 3 | 4 | 5 | 6 | 7 | 8 | 9 | 10 | 11 | 12 |
|---|---|---|---|---|---|---|---|---|---|---|---|---|
| 1. SSOR SI | – | | | | | | | | | | | |
| 2. Jacobi CG | <80 | – | | | | | | | | | | |
| 3. SOR | <80 | <80 | <80– | | | | | | | | | |
| 4. Sparse LU Uncomp + MD | <80 | <80 | <80 | – | | | | | | | | |
| 5. Sparse GE No PIV + MD | <80 | <80 | <80 | <80 | – | | | | | | | |
| 6. Sparse LU PIV + MD | 90 | <80 | <80 | <80 | <80 | – | | | | | | |
| 7. Sparse GE No PIV | 98 | 80 | <80 | <80 | <80 | <80 | – | | | | | |
| 8. Sparse LU Uncomp | 99 | 98 | 90 | <80 | <80 | <80 | <80 | – | | | | |
| 9. Envelope LUD | 99 | 99 | 99 | 80 | 80 | <80 | <80 | <80 | – | | | |
| 10. LINPACK Band | 99 | 99 | 99 | 80 | 80 | <80 | <80 | <80 | <80 | – | | |
| 11. Band GE No PIV | 99 | 99 | 99 | 90 | 80 | <80 | <80 | <80 | <80 | <80 | – | |
| 12. Sparse LU PIV | 99 | 99 | 99 | 99 | 99 | 90 | <80 | <80 | <80 | <80 | <80 | – |

**Table 5.4A.**  Pairwise confidence levels for the rankings of Table 5.4.

Table 5.5 indicates the ranking of 10 solvers for equations generated by General P2C1CL2. The pairwise confidence level of the rankings is given in Table 5.5A.

These data on confidence intervals produce groupings rather similar to these of the previous experiment (Tables 5.4 and 5.4A). Again, the iterative method is best followed by the three modules which are the minimum degree ordering. The orderings and performance of the worst four modules are very similar, except that Band GE No PIV and Sparse LU PIV trade positions. This should not be a great surprise, as the methods involved are quite similar.

### 5.4 Conclusions

The traditional and desirable way to present performance data is to state ''these data support hypothesis $H$ with confidence level $L$'' and to have explicitly stated hypotheses. Our data allows many such statements. We now give a few of them involving the direct comparison of particular pairs of solvers.

| Method | Average Rank | Median Slope | Sample run times for Prob. 1–1 | | |
|---|---|---|---|---|---|
| | | | 9×9 grid | 19×19 grid | 29×29 grid |
| SSOR SI | 1.00 | 1.9 | 0.9 | 2.6 | 6.1 |
| Sparse LU Uncomp + MD | 2.22 | 3.1 | 0.2 | 1.2 | 5.1 |
| Sparse LU PIV + MD | 3.44 | 3.2 | 0.2 | 1.3 | 6.2 |
| Sparse GE No PIV +MD | 3.89 | 3.2 | 0.1 | 1.4 | 6.0 |
| Sparse LU Uncomp | 5.44 | 3.3 | 0.2 | 1.7 | 7.0 |
| Sparse GE No PIV | 6.22 | 3.4 | 0.2 | 1.7 | 5.7 |
| Envelope LDU | 6.67 | 3.3 | 0.1 | 1.3 | 5.7 |
| Linpack Band | 7.78 | 3.4 | 0.2 | 2.4 | 10.7 |
| Sparse LU PIV | 8.89 | 3.5 | 0.2 | 2.2 | 10.1 |
| Band GE No PIV | 9.44 | 3.5 | 0.1 | 1.4 | 6.4 |

Table 5.5.   Average rankings and median slopes of 10 modules for solving the quadratic collocation equations (General P2C1CL2). Actual times for one problem are given in the right 3 columns. All timings are in seconds on a VAX 8600 using double precision.  seconds.

| | 1 | 2 | 3 | 4 | 5 | 6 | 7 | 8 | 9 | 10 |
|---|---|---|---|---|---|---|---|---|---|---|
| 1. SSOR SI | – | | | | | | | | | |
| 2. Sparse LU Uncomp + MD | <80 | – | | | | | | | | |
| 3. Envelope LDU PIV + MD | <80 | <80 | – | | | | | | | |
| 4. Sparse GE No PIV + MD | <80 | <80 | <80 | – | | | | | | |
| 5. Sparse LU Uncomp | 90 | <80 | <80 | <80 | – | | | | | |
| 6. Sparse GE No PIV | 99 | 80 | <80 | <80 | <80 | – | | | | |
| 7. Envelope LDU | 99 | 90 | <80 | <80 | <80 | <80 | – | | | |
| 8. Linpack Band | 99 | 99 | 90 | 80 | <80 | <80 | <80 | – | | |
| 9. Sparse LU PIV | 99 | 99 | 99 | 98 | <80 | <80 | <80 | <80 | – | |
| 10. Band GE No PIV | 99 | 99 | 99 | 99 | 80 | <80 | <80 | <80 | <80 | – |

Table 5.5A.   Pairwise confidence levels for the rankings of Table 5.5.

*Our data support the following with confidence level 99%:*

- SSOR SI has better slope than Band GE No PIV and LINPACK Band for all three experiments.

- The three sparse matrix, minimal degree modules have better slope than Band GE No PIV for the General P2C1CL2 equations.

- The Sparse LU Uncomp module has better slope than LINPACK Band for the General P2C1CL1 equations.

*Our data support the following with confidence level 98%:*

- The three sparse matrix, minimal degree modules have better slopes than Sparse LU PIV for the P2C1CL2 equations.

- Jacobi CG has better slope than Sparse LU Uncomp for the Interior P2C1CL2 equations.

*Our data support the following with confidence level 80%:*

- The modules Sparse LU Uncomp + MD and Sparse GE No PIV + MD have better slopes than any of the other non-sparse matrix modules for the Interior P2C1CL2 equations.

The data from a large experiment like this suggests many things, they are not established with high confidence or they are not even suitable for precise hypotheses or they need more data or different measurements. Examples of such subjective conclusions (better called conjectures or remarks) follow.

- LINPACK Band shows very poor performance for the General P2C1CL1 or P2C1CL2 equations.

- The iterative modules are dramatically less memory than the direct modules.

- Envelope LDU performs well for the P2C1CL1 equation, but not for the P2C1CL2 equations.

- The minimal degree indexing is effective for the P2C1CL2 equation (either General or Interior).

- The iterative modules have much smaller slopes than the other modules.

## REFERENCES

[1]   Boisvert, R.F., E.N. Houstis and J.R. Rice, A system for the performance evaluation of partial differential equations software, *IEEE Trans. Softw. Eng.*, **5** (1979), pp. 418–425.

[2]   Boisvert, R.F., Languages and software parts for elliptic boundary-value problems, *The role of languages in problem solving 2* (J.C. Boudreaux, B.W. Hamill and R. Jernigan, eds.), Elsevier Science Publishers (1987), pp. 411–431.

[3]   Bonomo, J., W.R. Dyksen and J.R. Rice, The ELLPACK performance evaluation system, Purdue University, Computer Science Department, Report CSD-TR 569 (1986), 23 pages.

[4]   Crowder, H., R.S. Dembo and J.M. Mulvey, On reporting computational experiments with mathematical software, *ACM Trans. Math. Software*, **5** (1979), pp. 193–203.

[5]   Dyksen, W.R., E.N. Houstis, R.E. Lynch and J.R. Rice, The performance of the collocation and Galerkin methods with Hermite Bicubics, *SIAM J. Numer. Anal.*, **21** (1984), pp. 695–715.

[6]   Hollander M. and D.A. Wolfe, *Non-parametric Statistics*, Chap.7, Wiley, New York (1973).

[7]   Houstis, E.N., C.C. Christara and J.R. Rice, Quadratic spline collocation methods for two point boundary value problems, *Int. J. Numerical Methods Eng.*, to appear.

[8]   Houstis, C.E., E.N. Houstis and J.R. Rice, Partitioning PDE computations: Methods and performance evaluation, *J. Parallel Comp.*, **4** (1987), pp. 141–163.

[9]   Houstis, E.N. and M. Irodotou-Ellina, An $O(h^6)$ quintic spline collocation method for fourth order two point boundary value problems, Purdue University, Computer Science Department, Report CSD-TR 616 (1986), 32 pages.

[10]  Houstis, E.N and J.R. Rice, High order methods for elliptic partial differential equations with singularities, *Int. J. Numerical Methods Eng.*, **18** (1982), pp. 737–754.

[11]  Houstis, E.N. and J.R. Rice, An experimental design for the computational evaluation of partial differential equation solvers, *The Production and Assessment of Numerical Software* (M. Delves and M.A. Hennell, eds.), Academic Press (1980), pp. 57–66.

[12] Houstis, E.N., E.A. Vavalis and J.R. Rice, Convergence of an $O(h^4)$ cubic spline collocation method for elliptic partial differential equations, *SIAM J. Numer. Anal.*, to appear.

[13] Lyness, J.N., Performance profiles and software evaluation, *Performance Evaluation of Numerical Software*, (L.D. Fosdick, ed.), North-Holland (1979), pp. 51–58.

[14] Rice, J.R., The algorithm selection problem, *Advances in Computers*, **15**, (Rubicoff and Yovits, eds.), Academic Press (1976), pp. 65–118.

[15] Rice, J.R, Methodology for the algorithm selection problem, *Performance Evaluation of Numerical Software*, (L. Fosdick, ed.), Horth-Holland (1979), pp. 301–307.

[16] Rice, J.R., Performance analysis of 13 methods to solve the Galerkin method equations, *J. Lin. Algebra Applica.*, **52/53** (1983), pp. 533–546.

[17] Rice, J.R., Design of a tensor product population of PDE problems, CSD-TR 628, Computer Science Department, Purdue University (1986), 12 pages.

[18] Rice, J.R. and R.F. Boisvert, *Solving Elliptic Problems with ELLPACK*, Springer-Verlag, New York (1985).

[19] Rice, J.R, E.N. Houstis and W.R. Dyksen, A population of linear second order, elliptic partial differential equations on rectangular domains, *Math. Comp.*, **36** (1981), pp. 475–484.

## APPENDIX: The PDEs

The partial differential operators used for this study are listed below. The domain for each problem is the unit square $0 \leq x, y \leq 1$, unless otherwise indicated. In each case, the forcing term $f(x, y)$ is determined to produce the specified true solution:

1–1    $Lu = (\exp(x*y)*u_x)_x + (\exp(-x*y)*u_y)_y - u/(1+x+y)$

      $u(x, y) = 0.75* \exp(x*y)* \sin(\pi*x)* \sin(\pi*y)$

4–1    $Lu = u_{xx} + u_{yy}$

      $u(x, y) = 3* \exp(x+y)* (x-x*x)* (y-y*y)$

5–1    $Lu = 4*u_{xx} + u_{yy} - a*u$

      $u(x, y) = 2* (x*x-x)* (\cos(2* \pi*y)-1)$

      $a = 0$

5–4    Operator 5–1 with $a = 10$

6–1    $Lu = u_{xx} + u_{yy} - (100 + \cos(2* \pi*x) + \sin(3* \pi*y))*u$

      $u(x, y) = -0.31*(5.4-C(x))*S(x)*(y*y-y)*(5.4-C(y))*(1/(1+F(x)**4)-.5)$

      $C(z) = \cos(4* \pi*z)$

      $S(z) = \sin(\pi*z)$

      $F(z) = 4* (z-.5)**2 + 4* (y-.5)**2$

7–1    $Lu = u_{xx} + u_{yy}$

      $u(x, y)$: approximate series

10–2   $Lu = u_{xx} + u_{yy}$

      $u(x, y) = \exp(-a* ((x-.5)**2 + (y-b)**2)) * (x*x-x)* (y*y-y)$

      $a = 50, \ b = .5$

10–3   Operator 10–2 with $a = 100, \ b = .5$

41–1   $Lu = u_{xx} + u_{yy} + a*u$     domain $[0,\pi]\times[0,\pi]$

      $u(x, y)$: approximate series (accuracy depends on $b$)

      $a = 10, \ b = 5.$

25–3   $Lu = -x**a*u_{xx} - y**a*u_{yy} - a*x** (a-1)*u_x - a*y** (a-1)*u_y + (x*y)**a$

      $u(x, y) = 3* \exp(x+y)* (x-x*x)* (y-y*y)$

      $a = 3.5$

# Scratchpad II: An Abstract Datatype System for Mathematical Computation

Richard D. Jenks, Robert S. Sutor and Stephen M. Watt
Computer Algebra Group
Mathematical Sciences Department
Thomas J. Watson Research Center
Yorktown Heights, NY 10598 USA

*Abstract*: Scratchpad II is an abstract datatype language and system that is under development in the Computer Algebra Group, Mathematical Sciences Department, at the IBM Thomas J. Watson Research Center. Many different kinds of computational objects and data structures are provided. Facilities for computation include symbolic integration, differentiation, factorization, solution of equations and linear algebra. Code economy and modularity is achieved by having polymorphic packages of functions that may create datatypes. The use of categories makes these facilities as general as possible.

## 1. Overview

Scratchpad II is

- an interactive language and system for mathematical computation

- a strongly-typed programming language for the formal description of algorithms, and

- a sophisticated tool kit for building libraries of interrelated abstract datatypes.

As an interactive system, Scratchpad II is designed to be used both by a naive user as a sophisticated desk-calculator and by an expert to perform sophisticated mathematical computations. Scratchpad II has very general capabilities for integration, differentiation, and solution of equations. In addition, it has an interactive programming capability which allows users to easily create new facilities or access those resident in the Scratchpad II library.

Scratchpad II is also a general-purpose programming language with a compiler used to add facilities to the system or user's library. Library programs are read by the system compiler, converted into object code, then loaded and executed through use of the system interpreter. The programming language and interactive language are identical except that library programs must be strongly typed. The unique abstract datatype design of Scratchpad II is based on the notion of *categories* and allows polymorphic algorithms to be expressed in their most natural setting and independently of the choice of data representation.

The Scratchpad II library consists of a set of parameterized modules (abstract datatypes) which collectively serve as a tool kit to build new facilities. Among these modules are those which create computational "types" (such as integers, polynomials, matrices and partial fractions) or data structures (such as lists, sets, strings, symbol tables, and balanced binary trees). These modules can be used to dynamically "mix and match" types to create *any* computational domain of choice, e.g. matrices of matrices, or matrices of polynomials with matrix coefficients.

In contrast with Scratchpad II, other existing computer algebra systems, such as MACSYMA, MAPLE, REDUCE and SMP use but a few internal representations to represent computational objects. To handle complicated objects, some of these systems overload the data structure for a canonical form (such as rational functions) and use flags to govern which coefficient and/or exponent domain is to be used. As more and more overloading is done to a single internal representation, programs become increasingly error prone and unmanageable. The complexity of systems designed in this way tend to grow exponentially with the number of extensions. The design approach of Scratchpad II has considerable advantages relative to these other systems with respect to modularity, extensibility, generality, and maintainability.

This paper introduces the reader to the language and concepts of Scratchpad II in a "bottom-up" manner, illustrating some interesting and varied interactive computations. Section 2 introduces the reader to the Scratchpad II language and interpreter. Sections 3-8 of the paper systematically introduce some of the more interesting types in the Scratchpad II world. Sections 9-11 highlight the facilities of the computer algebra library. Sections 12-15 then discuss the underlying high-level concepts of the language and system.

## 2. Preliminaries

In an interactive session with Scratchpad II, the interpreter reads input expressions from the user, evaluates the expression, then display a result back to the user. Input and output lines are numbered and saved in a history file. System commands to perform utilities such as reading files, editing, etc. are preceded by ")". Everything after "--" is a comment.

The following produces the same result as (5**2)+4.

```
5**2 + 4

   (1)   29
```

The previously computed expression is always available as the variable named %.

```
% + 1

   (2)   30
```

Large integer computations remain exact.

```
2**1000

   (3)
   1071508607186267320948425049060001810561404811705533607443750388370351051
   1124936122493198378815695858127594672917553146825187145285692314043598459
   7757469857480393456777482423098542107460506237114187795418215304647498358
   1941267398767555916554394607706291457119647768654216766042983165262438683
   7205668069376
```

Floating point numbers can be allowed to have many digits. Here is $\pi$ to 200 places.

```
precision 200

   (4)   200

numeric %pi

   (5)
   3.141 59265 35897 93238 46264 33832 79502 88419 71693 99375 10582 09749
   44592 30781 64062 86208 99862 80348 25342 11706 79821 48086 51328 23066
   47093 84460 95505 82231 72535 94081 28481 11745 02841 02701 93852 11055
   59644 62294 89549 30382
```

Symbols may be referenced before they are given values. It is easy to substitute something for the symbol at a later time.

```
(x + 11/111)**5

            5   555  4   123210  3   13676310  2   759035205      16850581551
   (6)   x  + --- x  + ------ x  + -------- x  + --------- x  + -----------
            11         121         1331           14641           161051
```

```
eval(%, x, 10)
```

$$(7) \quad \frac{1770223341829601}{16850581551}$$

## 3. Numbers

Scratchpad II provides many different kinds of numbers. Where appropriate, these can be combined in the same computation because the system knows how to convert between them automatically.

Integers can be as large as desired with the only limitation being the total storage available. They remain exact, no matter how large they get. Rational numbers are quotients of integers. Cancellation between numerators and denominators will occur automatically.

```
11**13 * 13**11 * 17**7 - 19**5 * 23**3 * 29**2
```

```
(1)   2538775111253891859464091805914957B
```

```
1/2 + 1/6 + 1/24 + 1/720 + 1/5040
```

$$(2) \quad \frac{1789}{2520}$$

For approximations, floating point calculations can be performed with any desired number of digits. The function *precision* sets the number of digits to use.

```
precision 39
```

```
(3)   39
```

A smaller precision might have given the impression that the following expression evaluated to 12. (Ramanujan wondered if it was actually an integer.)

```
numeric %pi * sqrt 310. / _   -- continued on next line
    log((2+sqrt 2.) * (3+sqrt 5.) * (5+2*sqrt 10.+sqrt(61+20*sqrt 10.))/4)
```

```
(4)   12.00 00000 00000 00000 00000 04945 80712 26995
```

Gaussian integers are complex numbers where both the real and imaginary parts are integers.

```
(5 + %i)**3
```

```
(5)   110 + 74%i
```

Of course, not all complex numbers have integer real and imaginary parts. The following number has floating point components.

```
(2.001 - 0.001 * %i)**2
```

```
(6)   4.004  - 0.004 002%i
```

Sometimes the form of a number is as important as the type of number. Here are a few ways of looking at integers and rationals in different forms.

```
factor 6432380707485569023720594412551704344145570763243
```

```
         13  11  7  5  3  2
(7)   11   13   17 19 23 29
```

```
continuedFraction(6543/210)
```

$$(8) \quad 31 + \cfrac{1}{6} + \cfrac{1}{2} + \cfrac{1}{1} + \cfrac{1}{3}$$

```
partialFraction(1,factorial(10))
```

$$(9) \quad \frac{159}{2^8} - \frac{23}{3^4} - \frac{12}{5^2} + \frac{1}{7}$$

-- now we expand the numerators into p-adic sums of the primes in the denominators

```
padicFraction %
```

$$(10) \quad \frac{1}{2} + \frac{1}{2^4} + \frac{1}{2^5} + \frac{1}{2^6} + \frac{1}{2^7} + \frac{1}{2^8} - \frac{2}{3^2} - \frac{1}{3^3} - \frac{2}{3^4} - \frac{2}{5^5} - \frac{2}{5^2} + \frac{1}{7}$$

We can also view rational numbers as radix expansions using various bases. Repeating sequences of digits are indicated by a horizontal line.

```
decimal(1/352)
```

$$(11) \quad 0.0028\overline{409}$$

```
base(4/7, 8)
```

$$(12) \quad 0.\overline{4}$$

Rational numbers raised to fractional powers can easily be created and manipulated.

```
(5 + sqrt 63 + sqrt 847)**(1/3)
```

$$(13) \quad \sqrt[3]{14\sqrt{7} + 5}$$

Integers modulo a given integer may be conveniently created and used.

```
123 mod 11      -- create an integer mod 11
```

$$(14) \quad 2$$

```
% + 79          -- operations involving this value are now done mod 11
```

$$(15) \quad 4$$

The following asserts that a is a number satisfying the equation $a^5 + a^3 + a^2 + 3 = 0$.

```
a | a**5+a**3+a**2+3 = 0
```

Among other things, this relationship implies that any expression involving a will never have it appear raised to a power greater than 4. We will define b so that it satisfies an equation involving a.

```
b | b**4+a = 0
```

```
2/(b-1)                     -- compute 2 times the inverse of (b-1)
```

$$(18)$$
$$(a^4 - a^3 + 2a^2 - a + 1)b^3 + (a^4 - a^3 + 2a^2 - a + 1)b^2$$
$$+$$
$$(a^4 - a^3 + 2a^2 - a + 1)b + a^4 - a^3 + 2a^2 - a + 1$$

```
2/%+1                       -- check result
```

$$(19) \quad b$$

There are many other varieties of numbers available, including cardinal numbers, which need not be finite, and quaternions, which are non-commutative.

```
Aleph 1 + Aleph 0
```

$$(20) \quad Aleph(1)$$

```
quatern(1,2,3,4)*quatern(5,6,7,8) - quatern(5,6,7,8)*quatern(1,2,3,4)
```

$$(21) \quad - 8i + 16j - 8k$$

## 4. Types

Every Scratchpad II object has an associated datatype, which can be thought of as a property of the object. The datatype determines the operations that are applicable to the object and cleanly separates the various kinds of objects in the system. If the user has issued

```
)set message type on
```

or, at least, has not turned it off,[1] the datatype of an object is printed on a line following the object itself. For example, if you enter 3.14159, the system will respond with a display similar to

$$(1) \quad 3.14159$$

```
Type: BF
```

In the Scratchpad II interpreter, BF is the abbreviation for BigFloat, which is the datatype of the number you entered. If you had not known anything about BF, issuing the command

```
)show BF
```

would have told you the unabbreviated name, the name of the file containing the Scratchpad II source code for BigFloat and the functions provided in the BigFloat *domain*.[2]

In the interpreter, each type has an abbreviation and it may be used almost anywhere the full name is used. Some of the abbreviations that are used in this paper are listed in Figure 1.

---

[1]   By default, it is on.

[2]   You can think of a *domain* as a collection of objects with a set of functions defined on the objects, plus a set of *attributes* that assert facts about the objects or the functions. For example, the domain Integer provides the integers, the usual functions on integers, and attributes asserting that multiplication is commutative, 1 is a multiplicative identity element, etc..

| Abbreviation | Full Name |
|---|---|
| A | Any |
| B | Boolean |
| BF | BigFloat |
| COMBINAT | CombinatoricFunctions |
| E | Expression |
| G | Gaussian |
| GF | GaloisField |
| I | Integer |
| L | List |
| P | Polynomial |
| QUEUE | Queue |
| RF | RationalFunction |
| RN | RationalNumber |
| S | String |
| SM | SquareMatrix |
| STK | Stack |
| ST | Stream |
| SY | Symbol |
| TBL | Table |
| UPS | UnivariatePowerSeries |

Figure 1.    Some Scratchpad II Type Names and their Abbreviations

In the previous section, each of the numbers really had a type, even though we chose not to display it. Some were simple, like Integer and BigFloat, and some were parametrized, like Gaussian Integer and ContinuedFraction Integer. Some of the types were fairly complicated, like SimpleAlgebraicExtension(RationalNumber, UnivariatePoly(x,RN), $a^{**}5 + a^{**}3 + a^{**}2 + 3$). At no point did we actually have to tell Scratchpad II the types of the objects we were manipulating. Although it is true that usually the Scratchpad II interpreter can determine a suitable type for an object without any type declarations whatsoever, you may sometimes want to supply additional information. You might provide this to help guide the interpreter to a particular type choice among several or to view an object in a particular way.

It is useful to know about types because:

1.   Scratchpad II really does use datatypes and they are present no matter how simple a model of the interpreter is discussed.

2.   Types are Scratchpad II objects in their own right and information is associated with them. A knowledge of types allows you to access and use this information.

3.   The use of explicit coercions with types provide a powerful way to transform an expression, be it to simplify the expression, change the output form, or to apply a particular function.

When you enter an expression in the Scratchpad II interpreter, the type inference facility attempts to determine the datatypes of the objects in the expression and to find the functions you have used. The following dialog demonstrates the types assigned by the interpreter to some simple objects.

```
23                  -- this is Integer

   (1)  23

Type: I
```

```
3.45                    -- this is BigFloat

    (2)  3.45

Type: BF

"this is a string"   -- this is String

    (3)  "this is a string"

Type: S

false                   -- this is Boolean

    (4)  false

Type: B

x                        -- this is Symbol

    (5)  x

Type: SY
```

The above expressions are *atomic*: they involve no function calls. When functions are present, things can get a bit trickier. For example, consider 2 / 3. By the basic analysis above, the interpreter determines that 2 and 3 belong to Integer. There is no function "/" in Integer so the interpreter has to look elsewhere for an applicable function. Among the possibilities are a "/" in RationalNumber that takes two elements of Integer and returns an element of RationalNumber. Since this involves no work in converting the arguments to anything else, this function is called and the rational number 2/3 is returned. This all happens automatically and is relatively transparent to the user.[3]

Associated with each type is a representation, a specific form for storing objects of the type. This representation is private and cannot be determined without examining the program which implements the type.

Some types, like Integer, are considered basic and have their representations provided internally by the system. Others, like RationalNumber, are built from other types (Record and Integer, here). Once a type is defined it may be used to represent other types. For example, QuotientField is represented by using Record and the type of the numerator and denominator. RationalFunction is represented by QuotientField Polynomial, along with the type of the coefficients of the polynomials. However, we re-emphasize that these details cannot be seen by users or other programs that manipulate values of these types.

Scratchpad II now provides over 160 different datatypes. Some of these clearly pertain to algebraic computational objects while others, like List and SymbolTable are data structures. Although Scratchpad II was originally designed as an abstract datatype language for computer algebra, no distinction is made to treat mathematical structures differently than data structures. In fact, data structures usually satisfy certain axioms and have mathematical properties of their own.

Scratchpad II is actually a general purpose language and environment: the new compiler for the language is being written in the language itself!

---

[3]    Some loading messages may appear from time to time as the system tries to coerce objects from one type to another or starts applying functions.

## 5. Lists

Lists are the simplest aggregate objects in Scratchpad II.

```
u := [1,4,3,5,3,6]

    (1)  [1,4,3,5,3,6]

rotate(u,2)

    (2)  [3,5,3,6,1,4]
```

Lists do not have to be homogeneous

```
u := [-43,"hi, there", 3.14]

    (3)  [- 43,"hi, there",3.14]
```

and they may be ragged.

```
v := [[1], [1,2,3], [1,2]]

    (4)  [[1],[1,2,3],[1,2]]
```

A monadic colon is used to append lists.

```
w := [:u, :[1..5],:u]    -- [1..5] is the list [1,2,3,4,5]

    (5)  [- 43,"hi, there",3.14,1,2,3,4,5,- 43,"hi, there",3.14]
```

Lists have origin 0. A "dot" is usually used to indicate indexing.

```
w.0

    (6)  - 43
```

Reduction over a list by a binary operator is supported.

```
*/[1..100]               -- this is 100 factorial

    (7)
9332621544394415268169923885626670049071596826438162146859296389521759999
9322991560894146397615651828625369792082722375825118521091686400000000000
00000000000000
```

A function may be applied to each element of a list by using "!".

```
oddp ! [1..5]            -- oddp returns true for an odd integer agrument

    (8) [true, false, true, false, true]

![1..5] + ![10..14]

    (9) [11, 13, 15, 17, 19]
```

A list may be viewed as a mapping which takes integers and returns the elements. The following list is then seen as the mapping

$$0 \to 1, 1 \to 1, 2 \to 2, ..., 7 \to 21.$$

```
u := [1,1..3,5,8,13,21]

   (10)  [1,1,2,3,5,8,13,21]
```

Juxtaposition with an intervening blank is equivalent to dyadic "." and means application. Parentheses are used for grouping. For lists, all three notations mean to apply the list as a mapping.

```
[u(0),u 1,u.2]

   (11)  [1,1,2]
```

A "!" can be used to apply *any* mapping to each element of a list.

```
u ! [0,1,3,5,7]

   (12)  [1,1,3,8,21]
```

Lists may created in many different ways. The following creates a list of the squares of the odd elements in *u*.

```
[n**2 for n in u | oddp n]

   (13)  [1,1,9,25,169,441]
```

A variety of very general iterator controls are available. Besides the "such that" form above, Scratchpad II also provides *while* and *until* forms. Iterations may also be nested or performed in parallel.

We now define a function *fib* to compute the Fibonacci numbers. The definition will be incrementally built from several separate pieces.

```
fib 0 == 1  -- the first initial value
fib 1 == 1  -- the second initial value
fib         -- looks at fib's value now as a mapping: 0 -> 1, 1 -> 1

   (16)  [1,1]
```

The general term will give a recursive definition for the remaining arguments of interest.

```
fib n==fib (n-1) + fib (n-2) when n > 1

fib         --look at its entire definition as a mapping

   (18)  [(n | 1 < n) -> fib(n - 1) + fib(n - 2),0 -> 1,1 -> 1]
```

The first term in the above mapping means if *fib* is given an argument n which is greater than 1, then fib(n) is computed using the recursive form. Now we will actually apply our function.

```
fib ! [0,1,3,5,7] --apply fib to each integer in our list of values
     compiling fib as a recurrence relation

   (19)  [1,1,3,8,21]
```

Note that we were able to determine that a recurrence relation was involved and specially compile the function.

## 6. Infinite Objects

Scratchpad II provides several kinds of infinite objects. We have already seen the example of a repeated decimal expansion of a rational number above. Other examples of infinite objects are streams and power series.

Streams are generalizations of lists which allow an infinite number of elements. Operationally, streams are much like lists. You can extract elements from them, use "!", and iterate over them in same way you do with lists.

There is one main difference between a list and stream: whereas all elements of a list are computed immediately, those of a stream are generally only computed on demand. Initially a user-determined number of elements of a stream are automatically calculated. This number is controlled by a *)set* user command and is 10 by default. Except for these initial values, an element of a stream will not be calculated until you ask for it.

The expression [n..] denotes the (primitive) stream of successive integers beginning with n. To see the infinite sequence of Fibonacci numbers, we apply *fib* to each member of [0..], the primitive stream of nonnegative integers.

```
fibs==fib![0..]
fibs              --by default, 10 values of a stream are computed

   (21)   [1,1,2,3,5,8,13,21,34,55,...]
```

Streams, like lists, are applicable as mappings and can be iterated over.

```
fibs ! [0,1,3,5,7]

   (22)   [1,1,3,8,21]

[n for n in fibs | oddp n]

   (23)   [1,1,3,5,13,21,55,89,233,377,...]

oddOnes s== [n for n in s | oddp n]  --define a function to do the filtering
oddFibs == oddOnes fibs              --define a new stream from the old

3*!oddFibs -1                        --produce [3*n-1 for n in oddFibs]

   (26)   [2,3,9,15,39,63,165,267,699,1131,...]

%![2*i for i in 1..]                 --can apply streams to streams

   (27)   [9,39,165,699,2961,12543,53133,225075,953433,4038807,...]
```

A power series can be obtained from a stream by coercing it to type UPS.

```
fibs::UPS(x,I)      --convert a stream to a power series

   (28)
```

$$1 + x + 2x^2 + 3x^3 + 5x^4 + 8x^5 + 13x^6 + 21x^7 + 34x^8 + 55x^9 + 89x^{10} + O(x^{11})$$

Another way to generate this power series is as follows:

1/ps(1-x-x**2)

(9)
$$1 + x + 2x^2 + 3x^3 + 5x^4 + 8x^5 + 13x^6 + 21x^7 + 34x^8 + 55x^9 + 89x^{10} + O(x^{11})$$

sin %        --the composition of one power series with another

(10)
$$x + 2x^2 + \frac{17}{6} x^3 + 4x^4 + \frac{541}{120} x^5 + \frac{13}{4} x^6 - \frac{15331}{5040} x^7 - \frac{3713}{180} x^8$$
$$+ \quad - \frac{22536359}{362880} x^9 - \frac{3046931}{20160} x^{10} + O(x^{11})$$

Power series can have coefficients from any ring, e.g. rational functions, gaussians, even other power series. Assuming m denotes a 2 x 2 square matrix with values 1,1,1,0, the following illustrates a power series with matrix coefficients.

1/ps(1-m*x)

(31)
$$\begin{bmatrix} 1 & 0 \\ 0 & 1 \end{bmatrix} + \begin{bmatrix} 1 & 1 \\ 1 & 0 \end{bmatrix} x + \begin{bmatrix} 2 & 1 \\ 1 & 1 \end{bmatrix} x^2 + \begin{bmatrix} 3 & 2 \\ 2 & 1 \end{bmatrix} x^3 + \begin{bmatrix} 5 & 3 \\ 3 & 2 \end{bmatrix} x^4$$
$$+ \begin{bmatrix} 8 & 5 \\ 5 & 3 \end{bmatrix} x^5 + \begin{bmatrix} 13 & 8 \\ 8 & 5 \end{bmatrix} x^6 + \begin{bmatrix} 21 & 13 \\ 13 & 8 \end{bmatrix} x^7 + \begin{bmatrix} 34 & 21 \\ 21 & 13 \end{bmatrix} x^8$$
$$+ \begin{bmatrix} 55 & 34 \\ 34 & 21 \end{bmatrix} x^9 + \begin{bmatrix} 89 & 55 \\ 55 & 34 \end{bmatrix} x^{10} + O(x^{11})$$

%::ST SM(2,I)    --obtain the coefficients of the power series as a stream

(32)
$$\left[ \begin{bmatrix} 1 & 0 \\ 0 & 1 \end{bmatrix}, \begin{bmatrix} 1 & 1 \\ 1 & 0 \end{bmatrix}, \begin{bmatrix} 2 & 1 \\ 1 & 1 \end{bmatrix}, \begin{bmatrix} 3 & 2 \\ 2 & 1 \end{bmatrix}, \begin{bmatrix} 5 & 3 \\ 3 & 2 \end{bmatrix}, \begin{bmatrix} 8 & 5 \\ 5 & 3 \end{bmatrix}, \right.$$
$$\left. \begin{bmatrix} 13 & 8 \\ 8 & 5 \end{bmatrix}, \begin{bmatrix} 21 & 13 \\ 13 & 8 \end{bmatrix}, \begin{bmatrix} 34 & 21 \\ 21 & 13 \end{bmatrix}, \begin{bmatrix} 55 & 34 \\ 34 & 21 \end{bmatrix}, \begin{bmatrix} 89 & 55 \\ 55 & 34 \end{bmatrix}, \ldots \right]$$

trace!%   --obtain a Fibonacci sequence, but with different initial conditions

(33) [2,1,3,4,7,11,18,29,47,76,...]

## 7. Functions

Functions can be as important as the values on which they act. In Scratchpad II functions are treated as first class objects; function-valued variables can be used in any way that variables of other types may be used.

Functions may be defined at top level, as were the maps from the previous section, or they may be obtained from a library of compiled code, as are the operations provided by types.

The simplest thing that can be done with a function object is to apply it to arguments to obtain a value.

```
5 + 6
```

```
    (1)   11
```

Type: I

If there are several functions with the same name, the interpreter will choose one of them. An attempt is made to choose the function according to certain generality criteria.

When a particular function is wanted, the plus on GF(7) for example, it can be specified by a *package call* using "$".

```
5 +$GF(7) 6
```

```
    (2)   4
```

Type: GF 7

Probably the next simplest thing is to assign a function value to a variable.

```
plusMod7  := _+$GF(7);  plusMod7(5, 6)  -- assigning + from GF(7) to a variable
```

```
    (3)   4
```

Type: GF 7

To access the value of the function object for a top level map it must be declared first.

```
double: I -> I
double n == 2*n
```

```
f := double;  f 13
```

```
    (6)   26
```

Type: I

Functions can be accepted as parameters or returned as values. Here we have an example of a function as a parameter

```
apply: (I -> I, I) -> I  -- apply takes a function as 1st parameter
apply(f, n) == f n       -- and invokes it on the 2nd parameter
```

```
apply(double, 32)
```

```
    (9)   64
```

Type: RN

and as a return value

```
trig: I -> (BF -> BF)    -- trig returns a function as its value
trig n == if oddp n then sin$BF else cos$BF
```

```
t := trig 1;  t 0.1
```

```
    (12)  0.099 83341 66468 28152 30681 4198
```

Type: BF

Several operations are provided to construct new functions from old. The most common method of combining functions is to compose them.

"*" is used for functional composition.

```
quadruple := double * double;  quadruple 3

   (13)  12

Type: I
```

"**" is used to iterate composition.

```
octuple := double**3;  octuple 3

   (14)  24

Type: I
```

*diag* gives the diagonal of a function.  That is, if g is diag f then g(a) is equal to f(a,a).

```
square := diag _*$I;  square 3

   (15)  9

Type: I
```

*twist* transposes the arguments of a function.  If g is defined as twist f then g(a,b) has the value f(b,a).

```
power := _**$RN;
rewop := twist power;  rewop(3, 2)

   (17)  8

Type: RN
```

Functions of lower arity can be defined by restricting arguments to constant values.  The operations *cur* and *cul* fix a constant argument on the right and on the left, respectively.  For unary functions, *cu* is used.

```
square := cur(power, 2);  square 4     -- square(a) = power(a,2)

   (18)  16

Type: RN
```

It is also possible to increase the arity of a function by providing additional arguments.  For example, *vur* makes a unary function trivially binary; the second argument is ignored.

```
binarySquare := vur(square);  binarySquare(1/2, 1/3)

          1
   (19)   -
          4

Type: RN
```

The primitive combinator for recursion is *recur*.  If g is recur(f) then g(n,x) is given by f(n,f(n-1,..f(1,x)..)).

```
fTimes := recur _*$NNI; factorial := cur(fTimes, 1::NNI); factorial 4
```

   (20)  24

Type: NNI

Functions can be members of aggregate data objects. Here we collect some in a list. The unary function incfn.i takes the i-th successor of its argument.

```
incfn := [(succ$SUCCPKG)**i for i in 0..5]; incfn.4 9
```

   (21)  13

Type: I

In practice, a function consists of two parts: a piece of program and an environment in which that program is executed. The display of function values appear as theMap(s, n), where s is a hideous internal symbol by which the program part of the function is known, and n is a numeric code to succinctly distinguish the environmental part of the function.

```
recipMod5 := recip$GF(5)
```

   (22)  theMap(MGF;recip;$U;17,642)

Type: GF 5 -> Union(GF 5,failed)

```
plusMod5  := _+$GF(5)
```

   (23)  theMap(MGF;+;3$;12,642)

Type: (GF 5,GF 5) -> GF 5

```
plusMod7  := _+$GF(7)
```

   (24)  theMap(MGF;+;3$;12,997)

Type: (GF 7,GF 7) -> GF 7

Notice above that the program part of plusMod5 is the same as for plusMod7 but that the environment parts are different. In this case the environment contains, among other things, the value of the modulus. The environment parts of recipMod5 and plusMod5 are the same.

When a given function is restricted to a constant argument, the value of the constant becomes part of the environment. In particular when the argument is a mutable object, closing over it yields a function with an *own* variable. For example, define shiftfib as a unary function which modifies its argument.

```
FibVals := Record(a0: I, a1: I)
```

   (25)  Record(a0: I,a1: I)

Type: DOMAIN

```
shiftfib: FibVals -> I
shiftfib r == (t := r.a0; r.a0 := r.a1; r.a1 := r.a1 + t; t)
```

Now fibs will be a nullary function with state. Since the parameter [0,1] has not been assigned to a variable it is only accessible by fibs.

```
fibs := cu(shiftfib, [0,1]$FibVals)
```

    (29)   theMap(%G12274,721)

Type: () -> I

```
[fibs() for i in 0..30]
```

    (30)
    [0, 1, 1, 2, 3, 5, 8, 13, 21, 34, 55, 89, 144, 233, 377, 610, 987,
    1597, 2584, 4181, 6765, 10946, 17711, 28657, 46368, 75025, 121393,
    196418, 317811, 514229, 832040]

Type: L I

## 8. Other Data Structures

We have seen that lists and streams can be used to hold values in a particular order.

```
[1980..1987, 1982, 1986]
```

    (1)   [1980,1981,1982,1983,1984,1985,1986,1987,1982,1986]

Scratchpad II provides many other structures that may better suit your applications. We will point out a few of them here.

Arrays provide a simple way of handling multi-dimensional collections of data.

```
a: Array([1..3,0..1,1..3], Symbol) := new s;
```

```
a(1,1,1) := a111;   a(1,0,2) := a102;   a(3,1,2) := a312;
```

```
a
```

$$(4) \quad \left[ \begin{array}{cc} s & a102 \\ a111 & s \end{array} \right., \left[ \begin{array}{cccc} s & s & s \\ s & s & s \end{array} \right., \left[ \begin{array}{cccc} s & s & s \\ s & a312 & s \end{array} \right] ]$$

Finite sets are collections of objects that contain no duplicates.

```
{1980..1987, 1982, 1986}
```

    (5)   {1980,1981,1982,1983,1984,1985,1986,1987}

A stack is a data structure where the last value added to it becomes the first one to be removed.

```
s : STK I := stack()
```

    (6)  stack(Bottom)

```
for i in 1980..1987 repeat push(i,s)
```

```
s
```

    (7)  stack(1987,1986,1985,1984,1983,1982,1981,1980,Bottom)

The value farthest from the bottom is the last one added.

```
pop s

   (8)   1987

s

   (9)   stack(1986,1985,1984,1983,1982,1981,1980,Bottom)
```

A queue is similar except that it is "first in, first out".

```
q : Queue I := queue()

   (10)   queue(Entry,Exit)

for i in 1980..1987 repeat enqueue(i,q)

q

   (12)   queue(Entry,1987,1986,1985,1984,1983,1982,1981,1980,Exit)

dequeue q

   (13)   1980

q

   (14)   queue(Entry,1987,1986,1985,1984,1983,1982,1981,Exit)
```

Scratchpad II provides several different types of tables to hold collections of values that can be looked up by some index set. The function *keys* gives a list of valid selectors to use to retrieve table entries.

Values of type Table(Key,Entry) are kept in memory in the workspace. Here Key and Entry may be replaced by any type.

```
colors : TBL(I, S) := table()

   (15)   table()

colors.1981 := "blue"; colors.1982 := "red"; colors.1983 := "green";

colors

   (17)   table(1981= "blue",1982= "red",1983= "green")

colors.1982

   (18)   "red"
```

KeyedAccessFile gives tables that are stored as random access files on disk. AssociationList is used for tables that may also be viewed as lists and have additional functions for looking up entries.

Record types are used to create objects with named components. The components of a record may be any type and do not all have to be the same type. An example declaration of a record is

```
bd : Record(name : S, birthdayMonth : I)
```

Here bd has two components: a String which is accessed via name and an Integer which has selector birthdayMonth.

You must set the value of the entire record at once if it does not already have a value. At this point is therefore illegal to enter bd.name := "Dick" because the birthdayMonth component has no value. However, bd := ["Dick", 11] is a legal assignment because it gives values to all components of the record. Issuing bd.name := "Chard" would now be legal.

A declaration such as

```
x : Union(I, S, BF)
```

states that x will have values that can be integers, strings or big floats. If, for example, the union object is an integer, the object is said to belong to Integer *branch* of the union.[4] The case infix operator returns a Boolean and can be use to determine the branch in which an object lies. The following function will display a message stating in which branch of the union the object x, defined above, lies.

```
sayBranch x ==
  if x case Integer then output "Integer branch"
  else if x case String then output "String branch"
  else if x case BigFloat then output "BigFloat branch"
  else output "don't know"
```

Now if we assign x := 8 and then issue

```
sayBranch x
```

```
   (3)   "Integer branch"
```

## 9. Algebraic Facilities

Scratchpad II provides a rich set of facilities for doing symbolic mathematical calculations. This section gives examples of integration, differentiation, solution of equations, and eigenvectors.

**Integration**

```
integrate(x**5/(x**4+x**2+1)**2,x)
```

$$
(7) \quad \frac{-x^2+1}{6x^4+6x^2+6} + \sum_{\alpha^2+\frac{1}{27}=0} \alpha * \log\left((x^2+2)\alpha + \frac{x^2}{3}\right)
$$

**Differentiation**

```
pderiv((x+1)*exp(log(x)/x+x**2/3)/(x-1),x)
```

$$
(33) \quad \frac{((-3x^2+3)\log(x)+2x^5-2x^3-3x^2-3)\%e^{\frac{3\log(x)+x^3}{3x}}}{3x^4-6x^3+3x^2}
$$

---

[4]    Note that we are being a bit careless with the language here. Technically, the type of x is always Union(I, S, BF). If it belongs to the Integer branch, x may be coerced to an object of type Integer.

```
integrate(%,x)      --check result
                         3
                   3log(x) + x
                   ------------
        x + 1           3x
(34)   (-----)%e
        x - 1
```

## Complex Zeros

```
solve(x**7+2*x**5-x**4+x**3-2*x**2-1=0,x,1/10000) --eqn, variable, precision
```

$$
(10) \quad [- \%i, \%i, -\ \frac{1}{2} - \frac{28377}{32768}\%i, -\ \frac{1}{2} + \frac{28377}{32768}\%i, 1]
$$

## Solution of Systems of Polynomial Equations

```
solve({x**2-x+2*w**2+2*y**2+2*z**2=0, 2*x*w+2*w*y+2*y*z-w=0,_
       2*x*y+w**2+2*w*z-y=0,x+2*w+2*y+2*z-1=0},_          -- set of equations
      {x,y,z,w}, _                                         -- set of variables
      1/1000)                                              -- precision
```

(6)

$$
[\{x=\frac{683}{2048}, y=0, z=\frac{683}{2048}, w=0\},\ \{x=1, y=0, z=0, u=0\},\ \{x=\frac{901}{2048}, y=\frac{215}{2048}, z=-\frac{271}{2048}, w=\frac{629}{2048}\},
$$

$$
\{x=\frac{1527}{2048}, y=-\frac{383}{2048}, z=\frac{165}{2048}, w=\frac{479}{2048}\},\ \{x=\frac{1157}{2048}, y=\frac{525}{2048}, z=-\frac{383}{2048}, w=\frac{305}{2048}\},
$$

$$
\{x=\frac{387}{2048}, y=\frac{155}{2048}, z=\frac{515}{2048}, w=\frac{161}{2048}\}]
$$

## Eigenvectors and Eigenvalues of a Matrix

```
eigenvectors [[x,2,1],[2,1,-2],[1,-2,x]]
```

(4)

$$
[[\text{eigval}= x + 1, \text{eigvec}= [\begin{bmatrix} 1 \\ 0 \\ 1 \end{bmatrix}]],
$$

$$
[\text{algrel}= (\%A - 1)x - \%A^2 + 9, \text{algvec}= [\begin{bmatrix} -1 \\ \dfrac{x - \%A - 1}{2} \\ 1 \end{bmatrix}]]]
$$

## 10. Coercion

Scratchpad II provides sophisticated facilities for changing an object of one type into an object of another type. If such a transformation involves no loss of information (as in creating a rational number from an integer), this process is called *coercion*. If some information may be lost (as in changing a rational number to a fixed precision floating point number), the process is called *conversion*. For the user, the major difference between coercions and conversions is that former may be automatically performed by the Scratchpad II interpreter while the latter must be explicitly requested.

The following is a definition of a function that computes Legendre polynomials.

```
leg(0) == 1
leg(1) == x
leg(n) == ((2*n-1)*x*leg(n-1)-(n-1)*leg(n-2))/n when n in 2..

leg 6
    Compiling function leg with signature I -> P RN
    Compiling function leg as a recurrence relation.
```

$$
(4) \quad \frac{231}{16} x^6 - \frac{315}{16} x^4 + \frac{105}{16} x^2 - \frac{5}{16}
$$

```
Type: P RN
```

From the expression on the right hand side of the definition of leg(n) the interpreter determined that the result type of the function should be Polynomial(RationalNumber). To see this result as a rational function with integer coefficients, just do a coercion.

```
% :: RF I
```

$$
(5) \quad \frac{231x^6 - 315x^4 + 105x^2 - 5}{16}
$$

```
Type: RF I
```

The double colon is the symbol for explicit coercion/conversion, where you are telling the interpreter, "I know what I want, so try to give me an object of this type."

As this example illustrates, coercion may be used to change the way an object looks. In this sense, coercion corresponds to the algebraic manipulation of formulas that one does, say, to simplify an expression or change it into a form that is more meaningful.

To illustrate this, let's start with a 2 by 2 matrix of polynomials whose coefficients are complex numbers. In this form, it doesn't make much sense to ask for the "real" part of the object. We will transform the matrix until we get a representation with a real and imaginary part, each of which is a matrix with polynomial coefficients. In the following, the symbol %i is the complex square root of 1. G is the abbreviation for Gaussian, a parameterized type used to create domains such as the complex numbers.

```
m : SM(2,P G I)

m := [[(j + %i)*x**k - (k + %i)*y**j for j in 1..2] for k in 1..2]
```

$$
(2) \quad \begin{bmatrix} (-1 - \%i)y + (1 + \%i)x & (-1 - \%i)y^2 + (2 + \%i)x \\ (-2 - \%i)y + (1 + \%i)x^2 & (-2 - \%i)y^2 + (2 + \%i)x^2 \end{bmatrix}
$$

```
Type: SM(2,P G I)
```

The matrix entries can be transformed so that they each have real and imaginary parts.

```
m :: SM(2, G P I)
```

$$
(3) \quad \begin{bmatrix} - y + x + (- y + x)\%i & - y^2 + 2x + (- y^2 + x)\%i \\ - 2y + x^2 + (- y + x^2)\%i & - 2y^2 + 2x^2 + (- y^2 + x^2)\%i \end{bmatrix}
$$

Type: SM(2,G P I)

Now we push the matrix structure inside the real and imaginary parts.

```
g := % :: G SM(2,P I)
```

$$
(4) \quad \begin{bmatrix} - y + x & - y^2 + 2x \\ - 2y + x^2 & - 2y^2 + 2x^2 \end{bmatrix} + \begin{bmatrix} - y + x & - y^2 + x \\ - y + x^2 & - y^2 + x^2 \end{bmatrix}\%i
$$

Type: G SM(2,P I)

It is now clearer what is meant by the "real part" of the object.

```
real(g)
```

$$
(5) \quad \begin{bmatrix} - y + x & - y^2 + 2x \\ - 2y + x^2 & - 2y^2 + 2x^2 \end{bmatrix}
$$

Type: SM(2,P I)

In fact, this is what would have been returned if you just asked for real(m). If we would rather see this last object as a polynomial with matrix coefficients, a simple coercion will do it.

```
% :: P SM(2,I)
```

$$
(6) \quad \begin{bmatrix} 0 & - 1 \\ 0 & - 2 \end{bmatrix} y^2 + \begin{bmatrix} - 1 & 0 \\ - 2 & 0 \end{bmatrix} y + \begin{bmatrix} 0 & 0 \\ 1 & 2 \end{bmatrix} x^2 + \begin{bmatrix} 1 & 2 \\ 0 & 0 \end{bmatrix} x
$$

Type: P SM(2,I)

## 11. Output

Besides to the character-oriented two-dimensional output you have already seen in this paper, Scratchpad II provides facilities for viewing output in FORTRAN format and in forms suitable for TeX[TMs] and the IBM Script Formula Formatter. The following equation is displayed in the standard Scratchpad II output format.

```
R = (2*x**2+4)**4/(x**2-2)**5
```

$$
(1) \quad R = \frac{16x^8 + 128x^6 + 384x^4 + 512x^2 + 256}{x^{10} - 10x^8 + 40x^6 - 80x^4 + 80x^2 - 32}
$$

---

[5]   TeX is a trademark of the American Mathematical Society.

The FORTRAN-style output of the equation is

```
R=(16*x**8+128*x**6+384*x**4+512*x**2+256)/(x**10 -10*x**8+40*x**6 -80*
*x**4+80*x**2 -32)
```

A form suitable for input to the TeX™ formula processor is

```
$$
{R={{{{16} \  {x \sp 8}}+{{128} \  {x \sp 6}}+{{384} \  {x \sp 4}}+{{
512} \  {x \sp 2}}+{256}} \over {{x \sp {10}} -{{10} \  {x \sp 8}}+{{40]
\  {x \sp 6}} -{{80} \  {x \sp 4}}+{{80} \  {x \sp 2}} -{32}}}}
$$
```

This is for input to the Script Formula Formatter:

```
:df.
<R=<<<16 % <x sup 8>>+<128 % <x sup 6>>+<384 % <x sup 4>>+<512 % <x sup
2>>+256> over <<x sup 10> -<10 % <x sup 8>>+<40 % <x sup 6>> -<80 % <x
sup 4>>+<80 % <x sup 2>> -32>>>
:edf.
```

When formatted by Script, the equation appears as

$$R = \frac{16\,x^8 + 128\,x^6 + 384\,x^4 + 512\,x^2 + 256}{x^{10} - 10\,x^8 + 40\,x^6 - 80\,x^4 + 80\,x^2 - 32}$$

The integration with respect to x of the right hand side of the equation produces a object which is a rational function plus a sum over the roots of a polynomial. The output produced by Scratchpad II for the Script Formula Formatter is

```
:df.
<<<-<10 % <x sup 7>> -<12 % <x sup 5>> -<24 % <x sup 3>> -<80 % x>>
over <<x sup 8> -<8 % <x sup 6>>+<24 % <x sup 4>> -<32 % <x sup 2>>+16>>
+<sum from <<< alpha  sup 2> -<9 over 2>>=0> of < alpha  % <log left {̲ <
<<x % alpha > -3>> right )>>>>
:edf.
```

The processed form is much easier to understand!

$$\frac{-10\,x^7 - 12\,x^5 - 24\,x^3 - 80\,x}{x^8 - 8\,x^6 + 24\,x^4 - 32\,x^2 + 16} + \sum_{\alpha^2 - \frac{9}{2} = 0} \alpha\,\log(x\,\alpha - 3)$$

## 12. Packages

In a large system there will be thousands of functions and there must be some way to organize them. One would be like to be able to group similar functions together and to be able to think in terms of useful collections of functions. In Scratchpad II, this is done with *packages*. For example, functions to compute permutations, combinations and partitions are be grouped together in a package providing simple combinatoric functions.

To see what functions are available in a package, the *show* system command is used.

```
)show CombinatoricFunctions

CombinatoricFunctions is a package constructor.

Abbreviation for CombinatoricFunctions is COMBINAT
Issue )edit ARITHMET SPAD to see source code for COMBINAT

---------------------- Operations --------------------------
binomial : (I,I) -> I          combination : (I,I) -> I
multinomial : (I,L I) -> I     partition : I -> I
permutation : (I,I) -> I       selection : (I,I) -> I
```

To group a collection of functions as a package, they must be compiled together in the body of a package constructor. A package constructor is a function which returns a Scratchpad II package object. This act of calling such a function is called *package instantiation.*

The package constructor for the CombinatoricFunctions is

```
CombinatoricFunctions(): T == B where
    T == with
         binomial:     (Integer,Integer) -> Integer
         multinomial:  (Integer, List Integer) -> Integer
         permutation:  (Integer,Integer) -> Integer
         combination:  (Integer,Integer) -> Integer
         selection:    (Integer,Integer) -> Integer
         partition:    Integer -> Integer

    B == add
         ArithmeticFunctions()  -- import factorial from another package

         binomial(n,k) ==
             k < 0 or n < k => 0
             k = 0 or n = k => 1
             n quo 2 < k     => binomial(n,n-k)
             t := 1
             for i in 1..k repeat t := (t*(n-i+1)) quo i
             t

                 ...
                 ...
         -- p is not exported, it is local to this package.
         p(m: Integer, n: Integer): Integer ==
             m = 1 => 1
             m < n => p(m-1,n) + p(m,n-m)
             m = n => p(m-1,n) + 1
             p(n,n)

         partition n == p(n,n)
```

This example serves to illustrate several points. The first line is the definition of the function CombinatoricFunctions which has type T and body B, with T and B defined further on. The type information for a package consists mainly of a list of the functions it exports and their types. The body gives the definitions of the exported functions. Because local variables in the body of the package constructor are invisible from outside, it is possible to maintain information which is private to the package.

## 13. Domains

One very natural way to group functions is to place together the operations for combining values of a given type. In one sense, the collection of operations which may be performed on values of a given type define what the type is. If these functions are provided by a single package, then it is possible to hide the representation of the values belonging to the type by keeping it local to the package. In Scratchpad II, using packages to so encapsulate a new types is the basic method of data abstraction.

For convenience we usually distinguish between packages which implement types and those which do not. We call the former *domains* and usually use the term *package* only for those which do not implement types.

We illustrate Stack below as an example of a domain constructor. The use of "$" in the signatures of exported operations (e.g. *pop*) represents the type which the domain implements.

```
Stack(S: Set): T == B where
    T == Set with
        stack:  ()-> $
        empty?: $ -> Boolean
        depth:  $ -> Integer
        push:   (S, $) -> S
        pop:    $ -> S
        peek:   $ -> S
        peek:   ($, Integer) -> S

    B == add
        -- Rep is a record so that the empty stack is mutable.
        Rep := Record(head: String, body: List S)

        Ex  ==> Expression
        coerce(s): Ex ==
            args: List Ex := []
            for e in s.body repeat args := cons(e::Ex, args)
            args := nreverse cons("Bottom"::Expression, args)
            mkNary("stack"::Ex, args)
        stack() ==
            ["Stack", []]
        empty? s ==
            null s.body
        push(e, s) ==
            s.body := cons(e, s.body)
            e
        pop s ==
            empty? s => error "Stack over popped."
            e := first s.body; s.body := rest s.body
            e
        peek s ==
            empty? s => error "Can't peek empty stack."
            first s.body
        depth s == #s.body
        peek(s,i) ==
            n := # s.body
            i > n-1 or i < -n => error "Out of bounds peek."
            s.body.(i<0 => n+i; i)
```

The coercion to Expression is used to give the output form of values in the domain.

## 14. Polymorphism

Whereas the package constructor for CombinatoricFunctions is a nullary function, in practice most package constructors take arguments as does Stack. Since package constructors may have type valued arguments, the exported functions may be used to express polymorphic algorithms.

The need for polymorphic functions stems from the desire to implement a given algorithm only once, and to be able to use the program for any values for which it makes sense. For example, the Euclidean algorithm can be used for values belonging to any type which is a Euclidean domain. The following package takes a Euclidean domain as a type parameter and exports the operations gcd and lcm on that type.

```
GCDpackage(R: EuclideanDomain): with
        gcd: (R, R) -> R
        lcm: (R, R) -> R
    == add
        gcd(x,y) ==                       -- Euclidean algorithm
            x:= unitNormal.x.coef
            y:= unitNormal.y.coef
            while y ¬= 0 repeat
                (x,y):= (y,x rem y)
                y:= unitNormal.y.coef
            x
        lcm(x, y) ==
            u: Union(R, "failed") := y exquo gcd(x,y)
            x * u::R
```

The exported operations are said to be *polymorphic* because they can equally well be used for many types, the integers or polynomials over GF(7) being two examples. Although the same *gcd* program is used in both cases, the operations it uses (*rem*, *unitNormal*, etc.) come from the type parameter R.

## 15. *Categories*

While polymorphic packages allow the implementation of algorithms in a general way, it is necessary to ensure that these algorithms may only be used in meaningful contexts. It would *not* be meaningful to try to use GCDpackage above with Stack(Integer) as the parameter. In order to restrict the use to cases where it makes sense Scratchpad II has the notion of *categories*.

A category in Scratchpad II is a restriction on the class of all domains. It specifies what operations a domain must support and certain properties the operations must satisfy. A category is created using a category constructor such as the one below.

```
OrderedSet(): Category == Set with
    -- operations
        "<": ($,$) -> Boolean
        max: ($,$) -> $
        min: ($,$) -> $
    -- attributes
        irreflexive "<"    -- not (x < x)
        transitive  "<"    -- x < y and y < z  implies  x < z
        total "<"          -- not(x < y) and not(y < x)  implies x=y
```

OrderedSet gives a category which extends the category Set by requiring three additional operations and three properties, or attributes.

A declaration is necessary in order for a domain to belong to a given category: having the necessary operations and attributes does not suffice. This is because the attributes are not intended to give a complete set of axioms, but merely to make explicit certain facts that may be queried later on. It is usually the case that belonging to a category implies that a domain must satisfy conditions that are not mentioned as attributes. For example, in OrderedSet there is no attribute relating *min* and " < ", although such relations are implied.

A type parameter to a domain or package constructor is usually required to belong to an appropriate category. For example, in the previous section the parameter R to GCDpackage was declared to belong to the category EuclideanDomain.

The use of categories in restricting the type parameters to a domain or package constructor allows algorithms to be specified in very general contexts. For example, since all table types belong to a common category, algorithms can be written that do not need to know the actual implementation (for example, whether it is a hash table in memory or a file on disk). As an example of algebraic generality, consider the domain of linear ordinary differential operators, which is declared as follows

```
LinearOrdinaryDifferentialOperator(A, M): T == B where
    A:  DifferentialRing
    M:  Module(A) with deriv: $ -> $

    T == GeneralPolynomialWithoutCommutativity(A,NonNegativeInteger) with
        D:      () -> $
        ".":    ($, M) -> M
    ...
    ...
```

This domain defines a ring of differential operators which act upon an A-module, where A is a differential ring. The type of the coefficients, A, is declared to belong to the category DifferentialRing and type of the operands, M, is declared to belong to the category Module(A) with a derivative operation. The constructed domain of operators is declared to belong to a category of general polynomials with coefficients A and two additional operations. The operation D creates a differential operator and "." provides the method of applying operators to elements of M.

It is often necessary to view a given domain as belonging to different categories at different times. Sometimes we want to think of the domain Integer as a belonging to Ring, sometimes as belonging to OrderedSet, and at other times as belonging to other categories. For a domain to have multiple views, it should be declared to belong to the Join of the appropriate categories. For example, the following keyed access file datatype may be viewed either as a table or as a file:

```
KeyedAccessFile(Entry: Set): public == private where
    FileRec  ==> Record(key: String, entry: Entry)
    ErrorMsg ==> String

    public == Join(FileCategory(LibraryName, FileRec),
            TableCategory(String, Entry, ErrorMsg)) with
            pack: $ -> $

    private == add ...
```

An important use of categories is to supply default implementations of operations. So long as certain primitive operations are provided by a domain, others can be implemented categorically. For example, supplying only " < " allows definitions of " > ", " < = " and " > = ". Thus a domain may inherit operations from a category. The use of Join provides multiple inheritance.

## Acknowledgments

The authors would like to thank Barry Trager, William Burge and Rüdiger Gebauer of the Computer Algebra Group at Yorktown Heights, and Greg Fee of the Symbolic Computation Group at the University of Waterloo for their suggestions and examples.

## Bibliography

[1]    Burge, W. H., and Watt, S. M., "Infinite Structures in Scratchpad II," *IBM Research Report* RC 12794 (Yorktown Heights, New York: May 27, 1987).

[2]    Computer Algebra Group, *Basic Algebraic Facilities of the Scratchpad II Computer Algebra System,* Yorktown Heights, New York: IBM Corporation, March 1936.

[3]    Jenks, R. D. and Trager, B. M., "A Language for Computational Algebra," *Proceedings of SYMSAC '81, 1981 Symposium on Symbolic and Algebraic Manipulation,* Snowbird, Utah, August, 1981. Also *SIGPLAN Notices,* New York: Association for Computing Machinery, November 1981, and *IBM Research Report* RC 8930 (Yorktown Heights, New York).

[4]     Jenks, R. D., "A Primer: 11 Keys to New Scratchpad," *Proceedings of EUROSAM '84, 1984 International Symposium on Symbolic and Algebraic Computation,* Cambridge, England, July 1984

[5]     Sutor, R. S., ed. *The Scratchpad II Newsletter*, Vol. 1, No. 1, Yorktown Heights, New York: IBM Corporation, September 1, 1985.

[6]     Sutor, R. S., ed. *The Scratchpad II Newsletter*, Vol. 1, No. 2, Yorktown Heights, New York: IBM Corporation, January 15, 1986.

[7]     Sutor, R. S., ed. *The Scratchpad II Newsletter*, Vol. 1, No. 3, Yorktown Heights, New York: IBM Corporation, May 15, 1986.

[8]     Sutor, R. S., and Jenks, R. D., "The Type Inference and Coercion Facilities in the Scratchpad II Interpreter," Proceedings of the SIGPLAN '87 Symposium on Interpreters and Interpretive Techniques, SIGPLAN Notices 22, 7, pp. 56-63, New York: Association for Computing Machinery, July 1987, and *IBM Research Report* RC 12595 (Yorktown Heights, New York: March 19, 1987).

# DATA PARALLEL PROGRAMMING AND BASIC LINEAR ALGEBRA SUBROUTINES

S. LENNART JOHNSSON[1]

**Abstract.** The Data Parallel programming model is conceptually simple and provides powerful programming primitives AS in shared memory models of computation. With an appropriate underlying architecture, primitives requiring global memory accesses do not require significantly longer execution times than primitives only requiring local accesses. In the data parallel programming model primitives are also available for expressing simple forms of local data interaction in relative coordinates, as for instance required in relaxation on multidimensional lattices.

The Connection Machine is a data parallel computer. In this paper we describe some of the data parallel aspects of the programming languages provided on the Connection Machine, comment upon the implementation of level-1 and level-2 BLAS, and describe the implementation of one level-3 BLAS function, matrix multiplication, in detail. For matrices of the size of the machine, or larger, only the kernel function is required. The kernel matrix multiplication yields a performance of up tp 5.2 Gflops in single precision on the CM-2 with the floating-point option. For matrices considerably smaller than the machine, all three nested loops in a Fortran 77 program can be parallelized, and expressed with a few instructions without any loop constructs.

**Key words.** Supercomputing, Parallel Computing, Basic Linear Algebra, Mathematical software.

**AMS(MOS) subject classifications.** 68Q10, 68Q25, 65F30

# 1   Introduction

High performance computation requires a high storage bandwidth. Supercomputers with a performance of a few Gflops need an effective storage bandwidth of the order of at least one Tbit/sec. With 10 MHz memory chips that is equivalent to a memory width of $10^5$ bits. If the problem has locality that can be exploited, then in a register or cache architecture the width of those facilities need to be $10^5$ bits, if operated at the same clock rate as the rest of the design. Even if locality is exploited the primary storage width needs to be substantial. We do not expect memory chips to become faster by more than a small constant factor. Hence, future supercomputers will require a storage with an even greater width. For most designs in VLSI technology the processor speed is comparable to, or a small constant factor higher than, the speed of a standard memory chip. With a machine designed entirely in VLSI technology the number of processors is a fraction of the number of memories in a system having a good balance between processing capability and storage bandwidth. The fraction is essentially determined by the width of the processor. A balanced design with supercomputer performance in todays technology would require in the order of 100,000 1-bit processors, or 3,000

---

[1]Thinking Machines Corp., 245 First Street, Cambridge, MA 02142, and Departments of Computer Science and Electrical Engineering, Yale University, New Haven, CT 06520

32-bit processors. With bandwidths in the order of a few Tbits/sec, or higher, it is increasingly difficult, and expensive, to have a bus for interconnecting memories with the processors. Interconnection networks are becoming common in parallel architectures. The Connection Machine described in some detail later has a total storage bandwidth of approximately 50 Gbytes/sec and $64k$ processors.

The programming of a computer with a large number of processors cannot be made in a way in which the programmer concerns himself with every processor, and their synchronization. Indeed, it is desirable that the programmer does not need to be concerned with details of the architecture, neither for functionality, nor for performance. Communication should largely be hidden from the programmer, as in sequential programming models, but clearly require efficient implementations. With a large number of processing elements and limited bandwidths at chip and board boundaries it is inevitable that the width of the communication channels be small. With a bit-serial pipelined communication system the time for global communication is of the same order as the time for local communication. Implementing a shared memory model of computation is a realistic proposition. Shared memory models of computation provide powerful operators, such as concurrent read, concurrent write with combining, and allows for the efficient implementation of parallel prefix operations.

In the data parallel programming model the emphasis is on the data structures, and the interaction between elements of the data structure. Language primitives may be provided for particular data structures and simple interactions within these structures. An example is multi-dimensional lattices, and nearest-neighbor communication in such structures. Operators for subsets are also provided, such as copying an element to every processor that needs that element, finding the maximum or minimum in a set, or the sum of all elements in a set. The global operators are quite powerful programming primitives.

The data parallel programming model is particularly suitable for highly concurrent architectures with a communications system that allows fast access to any part of the storage from any processor. Ideally there is one processor for every primary object of the problem. In reality that is rarely the case. Multiple elements have to be mapped to the same processor. The notion of *virtual processors* is used to allow the algorithm designer and the programmer to work with the data parallel model and not concern themselves with the mapping of the abstract machine to the real machine. The mapping of *virtual processors* to *real processors* is handled at compile time, if possible. The maximum problem size that can be handled is determined by the total amount of storage, not by the number of *real processors*, as in any realistic programming model. The storage of a *real processor* is divided up among *virtual processors*. The *virtual processors* assigned to a *real processor* time-share that processor. The scheme for assigning *virtual processors* to real processors affects the need for communication, and the *real processor* utilization. The most common assignment scheme is *consecutive* and *cyclic* [9]. For the matrix multiplication described in some detail later, either of these forms yield the same arithmetic and communication complexity, but a different control structure. The cyclic mapping is obtained by simply letting the low order bits of an address be the real processor address. In the *consecutive* mapping the high order bits are used for real processor addresses. The cyclic mapping yields a better load balance for LU-decomposition [9].

In the remainder of this paper we elaborate on the data parallel programming model, describe the Connection Machine architecture and some of the unique features of the Connection Machine programming languages, discuss implementation of level-1 and level-2 BLAS briefly, and treat in some detail a kernel for matrix multiplication, and the procedure for multiplying matrices of arbitrary shapes.

# 2  The Data Parallel programming model

In a data parallel programming model algorithms are devised, and described, for computing the solution with the structure of the problem domain in mind. The algorithms are expressed as a sequence of interactions between the elements of the structure. The type of interaction is often the same for large sets of elements. For instance, in solving partial differential equations a discretization of the continuous space is introduced by some approximation scheme, like finite differences, or finite elements. Then, an algorithm is devised for the solution of the resulting systems of equations. The solution can be expressed concisely as a sequence of interactions between the nodal quantities in the discrete space. The nodal quantities may be vectors, or single elements. The interactions may be local, or cover a large domain, potentially the entire domain. If an iterative method is used for the solution, then the interaction between elements of the data structure is described by the stencil used for the approximation of the differential operators, or by the type of elements used for finite element problems. If an elimination method is used for the solution of the system of equations, then the first several steps in a nested dissection procedure often only involves local interactions, but as the eliminations progress the interactions may extend over a larger domain. The more balanced the elimination tree, the further apart are typically the data elements involved in the final elimination stages. For a regular lattice it is of the order of one side of the lattice. In terms of the number of points involved it is of the order of $\sqrt{N}$ for a two-dimensional problem and $N^{\frac{2}{3}}$ for a three-dimensional problem.

Regardless of whether a direct or iterative technique is used for the solution of the system of equations, the type of interaction is in most cases the same for a very large set of the nodes in the discrete space. For instance, the same stencil is typically used in the entire domain, with the exception for the boundary. We have also implemented some high resolution schemes for shocks in fluid flow calculations, which use the same discretization over the entire domain. In adaptive techniques for the solution of partial differential equations, the same approximation may still be used in large regions of the domain. It may even be the case that only the resolution differs. In direct solvers the pivot row (column) interacts with every other row (column) for variables yet to be eliminated. The interaction is the same for every element. Even if the interaction is not identical for all elements, only a few types of interaction occurs in algorithms we are aware of. An essential characteristics of data parallel architectures is the ability to define classes of objects, and the operations thereupon.

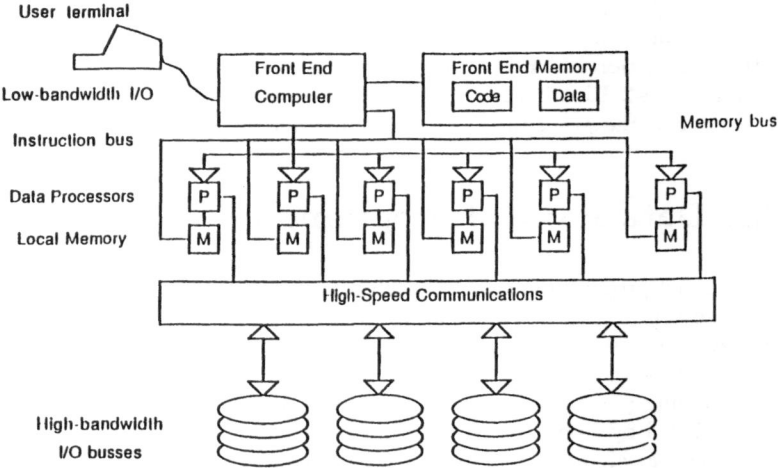

Figure 1: The Connection Machine system

# 3 The Connection Machine

## 3.1 Architecture

The Connection Machine is a data parallel architecture. It has a total primary storage of 512 *Mbytes* using 256 *kbit* memory chips, and 2 *Gbytes* using 1 *Mbit* memory chips. The bandwidth to storage is approximately 50 *Gbytes/sec*, which is achieved at a modest $8MHz$ clock rate. The primary storage has $64k$ ports, and a simple 1-bit processor for each port. The Connection Machine model CM-2 also has facilities for adding hardware for floating-point arithmetic. With the floating-point option a performance of 5.2 Gflops has been measured for matrix operations. We describe the algorithm used to realize this performance. This routine is the kernel for multiplying arbitrarily shaped matrices.

The Connection Machine needs a host computer. Currently, two families of host architectures are supported: the VAX family with the BI-bus, and the Symbolics 3600 series. The Connection Machine is mapped into the address space of the host. The program code resides in the storage of the host. It fetches the instructions, does the complete decoding of scalar instructions, and executes them. Instructions to be applied to variables in the Connection Machine are sent to a microcontroller, which decodes and executes instructions for the Connection Machine. The architecture is depicted in Figure 1

The difference between the nominal peak performance, and actual performance on the Connection Machine is largely due to the time spent in moving data between processors. Other overhead is low. The interprocessor communication capability depends on the communication pattern. For many linear algebra operations communication in a two-dimensional lattice is desirable, and broadcasting and summation along rows or columns. The lattice need not be a square lattice. The effective interprocessor communication bandwidth for two-dimensional lattice emulation is in the

| Algorithm | Arithmetic op's/ element | Arithmetic ops/ Elem. comm. |
|---|---|---|
| Matrix-vector mpy | 2 | $\sqrt{M}$ |
| Matrix-matrix mpy | $\frac{2}{3}\sqrt{N}$ | $\sqrt{M}$ |
| Relaxation 5-point stencil | 1.6/iter. | $\sqrt{M}$ |
| Radix-2 FFT | $1.25log_2N$ | $1.25log_2(M/2)$ |
| Sorting | $O(log_2N)$ - $O(log_2^2N)$ | |

Table 1: Number of operations per data element for a few operations on N elements in a machine with local storage M.

range $3 - 12$ *Gbytes/sec* for 4-byte messages. The overhead is approximately 15% for messages of this length. The communication time increases linearly with the message size. For communication in arbitrary patterns the Connection Machine is equipped with a router, which selects one of the shortest paths between source and destination, unless all of these paths are occupied. The router has several options for resolving contention for communication channels. The bandwidth for random permutations is a factor of 6 - 7 below that of the two-dimensional lattice emulation.

The degradation is due to bandwidth limitations at the chip boundaries. The performance measured as arithmetic or logic operations per unit time benefits from increased granularity for many operations in that the arithmetic/logic operations grows faster than the communication as a function of the number of local data elements, as shown in Table 1.

## 3.2 Programming languages

The Connection Machine programming languages are currently *Lisp, and a parallel version of C called C* is just becomming available. These languages are extensions of the familiar Lisp and C languages. The most essential extensions are the existence of a parallel data type, and the operations thereupon. Scans are among the operators included in the extensions. Concurrent read and concurrent write instructions are also supported. Members of a set of elements forming a parallel variable are operated upon concurrently by one instruction. No enumeration of the elements is required, and one or several loop levels disappears from the code, compared to languages not supporting array or set operations. The code becomes more compact, simpler, easier to debug, and one source of errors has vanished. Given that the programming languages for the Connection Machine are extensions of conventional languages the debugging tools, and the debugging process is similar to that for conventional architectures.

### 3.2.1 *Lisp

*Lisp is an extension of Common Lisp. There is one additional data type: a parallel variable, known as *pvar* . Parallel variables are defined through a (*defvar *pvar pvar-*

*expression* ) statement. In the current implementation a *pvar* is allocated across the entire configuration of the Connection Machine. The same section of the storage of every processor is assigned to a given *pvar* . The *pvar-expression* is optional. If present, a value is computed for every element of the *pvar* . Any element of a pvar can be referenced by the statement (**pref** *pvar address* ), which returns the element specified by *address* of the pvar specified by *pvar* . Assigning elements to pvars can be made through the function (**\*set** *pvar-1 pvar-2* ). The elements of *pvar-1* is set to the corresponding values of *pvar-2* . Individual elements of a pvar can be set by using the Lisp function setf : (setf (pref *pvar address* ) *var*)

\*Lisp provides two global addressing schemes: the conventional binary addresses, known as *cube-address* , and addresses in multi-dimensional lattices, *grid-address* . References on the grid addresses can be made both with absolute coordinates, and relative addresses. The forms are

(pref *pvar address* )
(pref-grid *pvar grid-address* )
(pref-grid-relative *pvar relative-grid-address* )

and for concurrent access

(pref !! *pvar-expression cube-address-pvar* )
(pref-grid !! *pvar-expression grid-address-pvars border-pvar* )
(pref-grid-relative !! *pvar-expression relative-grid-address-pvars border-pvar* )

where *grid-address* is the address of a lattice point, *relative-grid-address* the relative address of a lattice point. *grid-address-pvars* and *relative-grid-address-pvars* must contain as many address pvars as there are dimensions in the lattice. The *border-pvar* field is optional. If it is provided, then if a processor *p* references a processor outside the defined lattice, then instead the value of the pvar *border-pvar* in processor *p* is returned. The standard operators in Common Lisp are extended to pvars by the suffix !!. The following example illustrates the implementation of Jacobi iteration for the 5-point stencil

```
(*set new
    (*!!
        (!!0.25)
        (+!!
            (pref-grid-relative !! old (!! -1) (!! 0))
            (pref-grid-relative !! old (!! 0) (!! -1))
            (pref-grid-relative !! old (!! 0) (!! 1))
            (pref-grid-relative !! old (!! 1) (!! 0))
            rhs
        )
    )
)
```

The operation **\*set** is a local memory movement in all selected processors. The corresponding global operation is **\*pset** , which like **pref** comes in three forms depending on the addressing scheme.

(**\*pset** *combiner value-pvar destination-pvar cube-address* )
(**\*pset-grid** *combiner value-pvar destination-pvar grid-address-pvars* )
(**\*pset-grid-relative** *combiner value-pvar destination-pvar relative-grid-address-pvars* )

The content of *value-pvar* will be written into *destination-pvar* of the processor(s) specified by *cube-address* , *grid-address-pvars* , or *relative-grid-address-pvars* , respectively. Depending upon the addresses that are specified, several processors may be writing into the same memory location. The field *combiner* specifies the type of combining that is desired in such a case. Examples of *combiners* are **add**, **max**, and **min**.

In general, whether or not an operation shall be carried out is a function of the state. Examples of conditionals in \*Lisp are \*all, \*when, \*if, and \*cond.

**\*all** *body*
**\*when** *pvar body*
**\*if** *pvar-expression then-form else-form*
**\*cond** {(*pvar* {form}\*)}\*

In the \*all statement the body is evaluated for the entire set of processors. \*when subselects the processors for which *pvar* is non-NIL from the currently selected set. For the \*if statement the subset of processors of the currently selected set for which the *pvar* is non-NIL executes the *then-form*. The *else-form* is optional. \*cond evaluates *all* clauses. Subselection is based on the *pvar-expression* .

\*Lisp also has some powerful global operators. Of particular interest for numeric applications are \*min , \*max , and \*sum , and *scans*. The \*min , \*max , and \*sum operators computes the minimum, maximum, and the sum of the values in a pvar in the currently selected set of processors. Scans are specified by

(scan !! *pvar function* : direction *segment-pvar* :include-self)
(scan-grid !! *pvar function* :dimension : direction *segment-pvar* :include-self)

The *segment-pvar* divides the address space into non-overlapping segments. The scan operation is concurrently applied to all segments for which the *segment-pvar* is true. A scan is a parallel prefix operation, with the prefix specified by the *function* field. If the function is *plus*, then every pvar location will contain the sum of the pvar elements in processors with lower addresses in its segment, including its own original pvar value, if include-self is true. The scanning can also be made for decreasing addresses by specifying the direction to be backward. Note that for non-associative operators such as floating-point addition, precision may be lost.

## 3.2.2   C*

C* is an extension of C with a strong influence from C++. Objects that are of the same nature are in C* members of the same domain . Conceptually a domain is similar to a class in C++. In the data parallel model of computation a processor is associated with every instance of a domain . Every member of a domain has the same storage layout. Referencing a domain implies a selection of processors.

Only processors associated with an instance of the referenced domain remain active. There are two new data types: mono and poly . Data of type poly are allocated in the Connection Machine storage. Data belonging to any domain is by default of type poly . Data that is not of type poly is of type mono , and resides in the storage of the host machine.

There are only very few new operators in C*. Most C operators are extended to C* by the distinction between mono and poly . Communication between the host and the Connection Machine is implicit. Broadcasting from the host occurs if a mono value is assigned to a poly variable. Reduction occurs if the combined result of the elements of a poly variable is desired. The result of the reduction is a mono value in the host. Interaction between the elements of a single or several poly variables results in various communication patterns in the Connection Machine. C* does not currently allow the programmer to specify particular communication patterns, such as lattice communication.

### 3.2.3 *Fortran

The *Fortran that is being planned for the Connection Machine is similar to the proposed Fortan 8X standard. This standard has array constructs and operations on such constructs. In particular, the maxval, minval, and sum functions correspond to the *min , *max , and *sum functions already available in *Lisp, and will be supported in *Fortran. Much of the same comments as have been made for *Lisp and C* applies to *Fortran.

## 4   The BLAS

The level-1 BLAS are operations on a vector, or pairs of vectors. The vectors are pvars or poly variables in the data parallel model. Dot products are easily expressed in both *Lisp and C*. For instance, the *Lisp expression is *sum (*!! x y). The BLAS-1 copy routine is simply a *set . The level-1 BLAS routines do not require any loop constructs. The required data movement is implicit in the instructions, and the efficiency of level-1 BLAS a consequence of being at the level of machine instructions.

The level-2 BLAS routines can be written with two nested loops in Fortran 77. One of the operands is a matrix. With the machine configured as a two dimensional array the assignment of matrix elements to processors is simple. The level-2 BLAS routines have one or two vector operands. With the vectors aligned with one of the axis, a transposition may be (but need not be) required. The vector transposition is a single instruction. For matrix-vector multiplication one copy-scan-grid and one plus-scan-grid suffice in addition to a concurrent multiplication over all matrix elements. For a rank-1 update two copy-scan-grid instructions and the concurrent multiplication suffice. For the solution of a triangular system of equations loops can be avoided, if the inverse is represented instead of the triangular factors. The forward or backsolve then become matrix-vector multiplications. Otherwise, one loop is required.

Level-3 BLAS attempts to provide a standard interface for matrix-matrix operations. In the remainder of this paper we provide some insight into the techniques and considerations in implementing matrix multiplication on a data parallel architecture.

# 5 Multiplying arbitrarily shaped matrices

The computation we consider for the remainder of this paper is $A \leftarrow B \times C + D$, where the matrix $B$ is a $P \times Q$ matrix and $C$ a $Q \times R$ matrix and $A$ and $D$ $P \times R$ matrices. The particular issues we focus on are processor utilization for arbitrarily shaped matrices, effective use of the communication capabilities of an architecture such as the Connection Machine, and software engineering issues in terms of suitable primitives, in the spirit of the BLAS.

A general matrix computation can take several standard forms, such as an inner-product covered by the level-1 BLAS [16], an outer product or rank-1 update covered by the level-2 BLAS (GER) [4], an AXPY (level-1 BLAS) [16], or triad, or a matrix-vector product (GEMV) [4]. These are degenerate cases of matrix-matrix multiplication (MM), level-3 BLAS [3]. The non-degenerate matrix-matrix multiplication can be expressed in terms of the inner-product operation or the AXPY. The former is suitable for architectures with fast inner-product instructions, the latter for architectures with pipelined arithmetic units.

None of the BLAS versions directly addresses the issues in parallel computation. However, block algorithms that in part is the justification for introducing the level-3 BLAS work well on shared memory multiprocessors [8]. For distributed storage architectures a variety of systolic algorithms have been proposed for dense and banded matrices [1,2,15,17,10,12,11,6]. These algorithms defines the synchronization between the different data streams, and the alignment between those streams. The systolic algorithms are typically presented for one element per processor, but are easily generalized to a submatrix of each operand per processor [9,5]. For the systolic type algorithms constant storage suffice. A reduction in the communication time is possible in architectures with a high communications overhead, if data aggregation is allowed [13]. For a minimal communications time communication buffers and temporary storage of the order of $O(\sqrt{\frac{PQ}{N}})$ (or $O(\sqrt{\frac{PR}{N}})$, or $O(\frac{QR}{N})$) is necessary. In a data parallel architecture, such as the Connection Machine [7], the communications overhead is very low, and constant storage algorithms are preferable. The performance related issues are the utilization of the processors and the communication bandwidth for arbitrarily shaped matrices. From a software engineering point of view it is important to find programming/algorithm primitives that are useful for a variety of matrix shapes.

Matrix multiplication consists of three nested loops. In a data parallel architecture there is the potential for performing all three concurrently. In such a case the multiplication of a $P \times Q$ and a $Q \times R$ matrix can be performed in a time proportional to $log_2 Q$, if the number of processors $N \geq PQR$. Parallelizing the loop in the $Q$ direction is beneficial, if there are more than $PR$ processing elements. Our matrix multiplication algorithm for matrices of arbitrary shapes distinguishes between several cases. The kernel function that we describe here is based on an algorithm by Cannon [1]. This algorithm is devised for two-dimensional square lattices.

## 5.1 Data Allocation

For the kernel by Cannon the processor set is partitioned into a two-dimensional array of $N_r \times N_c = \lfloor \sqrt{N} \rfloor \times \frac{N}{\lfloor \sqrt{N} \rfloor}$ processors. For $log_2 N$ even the processing array is square, otherwise it is rectangular with one side twice the length of the other. For matrices such that $P, Q$ and $R$ are all smaller than the lattice dimensions one matrix element can be assigned to each processor. At the other extreme the number of matrix elements in all dimensions may exceed the number of processors in that dimension. Multiple elements have to be assigned to every processor, with a strategy to minimize the maximum storage per processor. Two apparent schemes are *cyclic* and *consecutive* assignment [9]. The two assignment schemes can be illustrated as follows, assuming $P = 2^p$, $Q = 2^q$, and $R = 2^r$ for simplicity.

$$\big(\underbrace{u_{p-1}u_{p-2}\ldots u_{n_r}}_{vp}\;\underbrace{u_{n_r-1}u_{n_r-2}\ldots u_0}_{rp}\;\underbrace{v_{q-1}v_{q-2}\ldots v_{n_c}}_{vp}\;\underbrace{v_{n_c-1}v_{n_c-2}\ldots v_0}_{rp}\big)$$

$$\big(\underbrace{u_{p-1}u_{p-2}\ldots u_{p-n_r}}_{rp}\;\underbrace{u_{p-n_r-1}u_{p-n_r-2}\ldots u_0}_{vp}\;\underbrace{v_{q-1}v_{q-2}\ldots v_{q-n_c}}_{rp}\;\underbrace{v_{q-n_c-1}v_{q-n_c-2}\ldots v_0}_{vp}\big)$$

The real address field is made up of $n_r + n_c = n$ address bits obtained by concatenating the real processor row address field with the real processor column address field.

In the *consecutive* assignment each processor is assigned a block matrix of size $2^{p-n_r} \times 2^{q-n_c}$ of matrix $B$, a block matrix of size $2^{q-n_r} \times 2^{r-n_c}$ of matrix $C$, and a block matrix of size $2^{p-n_r} \times 2^{r-n_c}$ of matrix $A$. The low order bits are used for *virtual processor* addresses of both the row and the column address fields. In the *cyclic* assignment the low order bits are instead used for *real processor* addresses, and high order bits for *virtual processors*. Mixed assignment schemes can be used as well [14], but will not be considered here. In the cyclic assignment the matrices are "tiled" with tiles of size $2^{n_r} \times 2^{n_c}$. Each such tile represents a slice of storage across all processors [9].

## 5.2 A Kernel for Concurrent Matrix Multiplication

In the algorithm by Cannon the inner products defining the elements of $A$ are accumulated *in-place*. Denote the storage cells for $A, B, C$ and $D$ by $E, F$ and $G$. The algorithm has two phases. A set-up phase in which the operands are aligned, and a multiplication phase. In the set-up phase the shifting yields: $F(i,j) \leftarrow F(i,(i+j) \pmod{2^k})$, $G(i,j) \leftarrow G((i+j) \pmod{2^k},j)$, $E \leftarrow D$ for $(i,j) \in \{0,1,2,...,2^k-1\} \times \{0,1,2,...,2^k-1\}$. Clearly $F(i,j) \times G(i,j)$ is a valid product for all $i$ and $j$. In the multiplication phase the following operations are carried out: $E(i,j) \leftarrow E(i,j) + F(i,j) \times G(i,j)$, $F(i,j) \leftarrow F(i,(j+1) \pmod{2^k})$, $G(i,j) \leftarrow G((i+1) \pmod{2^k},j)$, $i,j = \{0,1,2,...,2^k-1\}$.

We distinguish between the following cases:

1. The operands covers the processor array exactly.

2. The number of elements in each of the two dimensions of the operands are multiples of the number of processing elements in the corresponding dimension.

3. For each of the matrices the number of elements in at least one dimension is less than the number of processors assigned to that dimension.

## 5.2.1 Matrices Perfectly Matching the Mesh Size

The first case corresponds to the kernel function for $N_r = N_c$. If the number of processors is an odd power of two then a mesh with aspect ratio two is the closest to a square mesh. With square matrices each real processor has to simulate two virtual processors. Note that with the same storage scheme for all matrices and no matrix dimension exceeding the processor mesh $Q = min(N_r, N_c)$. Hence, either $P = max(N_r, N_c)$ and $R = min(N_r, N_c)$ or vice versa. The multiplication can be performed by first replicating the matrix $C$ in the row direction $\frac{max(N_r, N_c)}{min(N_r, N_c)}$ times if $P > R$, else replicating $B$ if $R > P$. The kernel algorithm is then applied concurrently to the $\frac{max(N_r, N_c)}{min(N_r, N_c)}$ meshes of size $min(N_r, N_c) \times min(N_r, N_c)$.

## 5.2.2 Virtual processors

In the second case each *real processor* has several *virtual processors*. In the *consecutive* assignment each *real processor* has a block matrix that with matrix dimensions being powers of two are of size $2^{p-n_r} \times 2^{q-n_c}$ for matrix $B$, of size $2^{q-n_r} \times 2^{r-n_c}$ for matrix $C$, and of size $2^{p-n_r} \times 2^{r-n_c}$ for matrix $A$. For the case of $N_r = N_c = \frac{n}{2}$ each *real processor* performs a block matrix operation requiring $2 \cdot 2^{p+r+q-\frac{3}{2}n}$. operations. Then, a rotation of the matrices $B$ and $C$ is performed blockwise, and the process repeated $2^{\frac{1}{2}n}$ times. The time for arithmetic is $\frac{PQR}{N}$, and the data transfer time proportional to $\frac{(P+R)Q}{\sqrt{N}}$. In the bit-serial pipelined communication system of the Connection Machine the overhead in the communication is negligible. If $N_r \neq N_c$, then with $N_r > N_c$ $2^{n_r-n_c}$ rotations are performed in the direction of $N_r$ for every rotation in the $N_c$ direction. For $N_r < N_c$ the situation is the opposite. The arithmetic time is the same as in the case of a square mesh, but the communication time is $(\frac{P}{N_c} + \frac{R}{N_r})Q$.

In the case of the *cyclic* assignment it is clear that every real processor has the same number of virtual processors as with *consecutive* assignment. It can also be shown that the amount of work for each communication is the same. In addition to performing the work for each virtual processor, data for different virtual processors are also in the correct location for a multiply-add operation. The control structure is different from the *consecutive* assignment, however.

For the local matrix multiplication the issues with respect to performance are the traditional ones, with block methods being preferable for cache based architectures, and AXPY based algorithms preferable for pipelined, and register architectures. The performance of our implementation of Cannon's algorithm extended to handle virtual processors is 5.2 Gflops with 256 virtual processors per real processor (or 4k x 4k matrices).

| $S_r$ | no. of block copy | no. of block additions |
|---|---|---|
| $P_r$ | $\frac{R_r}{P_r}$ | $\frac{Q_r}{P_r}$ |
| $Q_r$ | $\frac{R_r+P_r}{Q_r}$ | none |
| $R_r$ | $\frac{P_r}{R_r}$ | $\frac{Q_r}{R_r}$ |

Table 2: Number of block copy and addition for one product matrix.

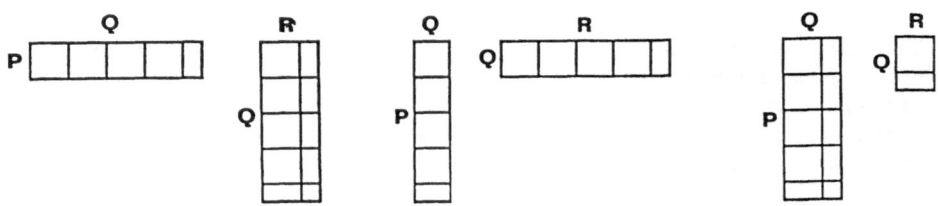

Figure 2: Multiplying arbitrary matrices on a large array

### 5.2.3 Small Matrices in a Large Mesh

This case is relevant not only for matrices with fewer elements than the number of processors, but also handles the case when $P$, $Q$, or $R$ are not multiples of the number of processors in the respective direction. In the *cyclic* assignment there are $PmodN_r$ rows in the last row of "tiles" of $B$ with rows being allocated in the direction of $N_r$. Similarly, there are $QmodN_c$ columns in the last columns of "tiles" of $B$. In the *consecutive* partitioning some processors have one more matrix element than others. Let $PmodN_r = P_r$, $QmodN_r = Q_r$, $QmodN_c = Q_c$, and $RmodN_c = R_r$. Assume $N_r = N_c$. Since $P_r, Q_r, R_r < N_r$ there are more processors than matrix elements, and there is a possibility to parallelize the loop in the $Q$ direction.

The operands are partitioned into squares of side $S_r = min(P_r, Q_r, R_r)$. There is a total of $\lceil \frac{P}{S_r} \rceil \lceil \frac{Q}{S_r} \rceil \lceil \frac{R}{S_r} \rceil$ such squares. Each such processing square receives a $S_r \times S_r$ block from $B$ and $C$ through a copy operation. The number of block copy and additions are summarized in Table 2.

All multiplications on blocks of size $S_r \times S_r$ are performed concurrently. Each block matrix multiplication use the kernel routine and requires $S_r$ multiplication steps. In general, a summation of corresponding elements in some of the blocks are necessary to complete the computation. For instance, if $P_r = 1, R_r = 1$ and $Q_r > 1$, then the computation is an inner-product. The blocks are of size $1 \times 1$ and a summation of the product of all blocks is required. Figure 2 illustrates the three different situations summarized in Table 2.

The number of processors used in the block algorithms is $\frac{PQR}{S_r}$. If the total number of processors can be partioned into several such sets of subarrays, say $M$, then the loop on $Q_r$ can be subdivided further at most that many times. The loop on $Q_r$ is parallelized in the algorithms above, except if $S_r = Q_r$. This further parallelization of the multiplication is obtained by creating multiple instances of the algorithms just described by copying, and by rotating the different copies $k\frac{S_r}{M}$, $k = \{1, 2, \ldots, M-1\}$ steps with respect to the "original". This rotation can be done while copying. The multiplication then proceeds in $\frac{S_r}{M}$ steps, followed by block addition of all copies.

If $N_r \neq N_c$, then $Q_r$ and $Q_c$ may not be the same. However, the number of rotations required for the direction with the larger number of processors is higher. For every rotation in the direction of $min(N_r, N_c)$, $\frac{max(N_r, N_c)}{min(N_r, N_c)}$ rotations are required in the other direction. This in effect guarantees that after the phase where all "complete" layers are treated for the direction with the maximum number of layers, the remainders in the two directions are the same.

# 6 Summary

The Data Parallel programming model provide a simple conceptual framework for programming highly concurrent architectures. The model also provide powerful instructions. Indeed, several of the level-1 BLAS routines are single instructions. None of the routines require loop constructs. The level-2 BLAS require two or three instructions, with the exception of the triangular system solvers, no loop constructs are required. If the inverse of the triangular matrix is available instead of the matrix, then the triangular solve can be accomplished in one or two instructions. Otherwise, a single loop is required.

Level-3 BLAS require three nested loops in Fortran 77. In the data parallel programming model a single loop is required, in general. Indeed, matrix multiplication can be performed without loops, if there is a sufficiently large number of processors. In this paper we use a simple mesh algorithm as the kernel. This kernel requires a single loop, and the communication operation is single step rotations, ignoring the initial data alignment. The kernel executes at a rate of 5.2 Gflops on the Connection Machine model CM-2. The loop is in the $Q$ direction for multiplying a $P \times Q$ and a $Q \times R$ matrix. This loop can be parallelized given that there is a sufficient number of processors. Parallelizing this loop implies the introduction of segmented copy-scans and plus-scans. For sufficiently many processors matrix multiplication can be performed without loops in the data parallel model of computing.

# References

[1] L.E. Cannon. *A Cellular Computer to Implement the Kalman Filter Algorithm.* PhD thesis, Montana State University, 1969.

[2] E. Dekel, D. Nassimi, and Sartaj Sahni. Parallel matrix and graph algorithms. *SIAM J. Computing*, 10:657–673, 1981.

[3] Jack J. Dongarra, Jeremy Du Croz, Iain Duff, and Sven Hammarling. *Preliminary Proposal for a Set of Level 3 BLAS.* Technical Report Technical Memorandum, Argonne National Laboratories, Mathematics and Computer Science Division, January 1987.

[4] Jack J. Dongarra, Jeremy Du Croz, Sven Hammarling, and Richard J. Hanson. *An Extended Set of Fortran Basic Linear Algebra Subprograms*. Technical Report Technical Memorandum 41, Argonne National Laboratories, Mathematics and Computer Science Division, November 1986.

[5] Geoffrey C. Fox, S.W. Otto, and A.J.G. Hey. *Matrix Algorithms on a Hypercube I: Matrix Multiplication*. Technical Report Caltech Concurrent Computation Project Memo 206, California Institute of Technology, dept. of Theoretical Physics, October 1985.

[6] Donald E. Heller. *Partitioning Big Matrices for Small Systolic Arrays*, pages 185–199. Prentice-Hall, 1985.

[7] W. Daniel Hillis. *The Connection Machine*. MIT Press, 1985.

[8] William Jalby and Ulrike Meier. *Optimizing Matrix Operations on a Parallel Multiprocessor with a Memory Hierarchy*. Technical Report , Univ. of Illinois, Center for Supercomputer Research and Development, February 1986.

[9] S. Lennart Johnsson. Communication efficient basic linear algebra computations on hypercube architectures. *Journal of Parallel and Distributed Computing*, 4(2):133–172, April 1987. (Report YALEU/CSD/RR-361, January 1985).

[10] S. Lennart Johnsson. *Computational Arrays for Band Matrix Equations*. Technical Report 4287:TR:81, Computer Science, California Institute of Technology, May 1981.

[11] S. Lennart Johnsson. Highly concurrent algorithms for solving linear systems of equations. In *Elliptic Problem Solving II*, Academic Press, 1983.

[12] S. Lennart Johnsson. Vlsi algorithms for doolittle's, crout's and cholesky's methods. In *International Conference on Circuits and Computers 1982, ICCC82*, pages 372–377, IEEE, Computer Society, September 1982.

[13] S. Lennart Johnsson and Ching-Tien Ho. Matrix multiplication on boolean cubes using generic communication primitives. In *Parallel Processing and Medium Scale Multiprocessors*, SIAM, 1987. (Report YALEU/CSD/RR-530, March 1987).

[14] S. Lennart Johnsson and Ching-Tien Ho. Matrix transposition on boolean n-cube configured ensemble architectures. *SIAM J. on Algebraic and Discrete Methods*, . To appear.

[15] H.T. Kung and Charles E. Leiserson. *Algorithms for VLSI Processor Arrays*, pages 271–292. Addison-Wesley, 1980.

[16] C. L. Lawson, R.J. Hanson, D.R. Kincaid, and F.T. Krogh. Basic linear algebra subprograms for fortran usage. *ACM TOMS*, 5(3):308–323, September 1979.

[17] Uri Weiser and Al Davis. *Mathematical Representation for VLSI Arrays*. Technical Report UUCS-80-111, Universit of Utah, Department of Computer Science, September 1980.

# INTEGRATING SYMBOLIC, NUMERIC AND GRAPHICS COMPUTING TECHNIQUES

## PAUL S. WANG*

**Abstract** A workstation-based integrated scientific computing system is a powerful tool for contemporary scientists and engineers. When symbolic, numeric and graphics computing techniques are integrated, a power environment for scientific computing results. The integrated tool is much greater than the sum of its constituent parts. We present some recent developments in this direction: (1) symbolic derivation of numerical code for finite element analysis; (2) automatic numeric code generation based on derived formulas; (3) interactive graphing of curves and surfaces for mathematical formulas; and (4) graphical user interface for mathematical systems. Prototype software systems in these directions are described with actual machine generated examples.

1. **Introduction.** Symbolic computation systems specialize in the exact computation with numbers, symbols, formulas, vectors, matrices and the like. When a problem is solved symbolically, the answers are exact mathematical expressions and there is no loss of precision. In some cases, the expressions obtained can provide a great deal of insight. However, many practical problems do not have neat exact solutions. Numerical packages, on the other hand, use floating-point numbers, and iterated approximate computations to solve problems. Many efficient numeric algorithms have been developed and programmed to solve a wide variety of scientific and engineering problems. A principal difficulty with numeric computation is loss of precision. Sometimes estimating the error in the final numerical solution is still a problem. It is usually the case that a problem has some parts that are suitable for symbolic processing and other parts that are better solved numerically. Thus, the two computational approaches are complementary and should be made available in an integrated scientific computing environment. In addition such an environment must also have adequate graphics facilities. Graphics can be used to convey information visually and in many situations is much more effective than formulas or tables of numbers. Obvious applications include the display and manipulation of points, curves and surfaces. Graphics is also important for advanced user interface design which is an important factor of whether a powerful, multi-functional system is easy to learn and use. The combination of symbolic, numeric and graphics computing techniques in a single

*Department of Mathematical Sciences, Kent State University, Kent, Ohio, 44242; Work reported herein has been supported in part by the National Science Foundation under Grant DCR-8504824, and in part by the Department of Energy under Grant DE-AC02-ER7602075-A013.

scientific computing environment can bring not only convenience but also new approaches for problem solving.

Modern workstations make it feasible to investigate such integrated computing environments. A workstation-based integrated scientific system should be the tool of choice for contemporary scientists and engineers. It is relatively simple to bring numeric, symbolic and graphics computing capabilities to a single computing system. What is more difficult is to have a truly integrated system where these techniques work together with very little barrier between them. More importantly, these three techniques should reinforce one another so that the whole is bigger than the sum of the parts. We present some recent developments in this direction:

1. symbolic derivation of numerical code for finite element analysis;

2. automatic numeric code generation based on derived formulas;

3. interactive graphing of curves and surfaces for mathematical formulas; and

4. graphical user interface for mathematical systems.

We shall also describe some software packages being developed in these areas.

2. **Symbolic derivation of finite element code.** Finite element analysis, has many applications in structural mechanics, heat transfer, fluid flow, electric fields and other engineering areas. It plays a vital role in modern Computer Aided Design. Large numerical packages such as NFAP [1] and NASTRAN [13] exist for finite element analysis. They provide facilities for frequently used models and cases. Only slight modifications of the "canned" computational approaches are allowed via parameter setting. Without extensive reprogramming of the formulas involved, these "canned" packages can not be used in situations where new formulations, new materials or new solution procedures are required. We have implemented a prototype software system to automate the derivation of formulas in finite element analysis and the generation of programs for the numerical calculation of these formulas. Some previous work in this area can be found in [4] and [5]. The promise and potential benefit of such an approach are clearly indicated. However, it is not enough for the approach to work well on simple problems that are limited in size and complexity. Practical problems in finite-element analysis involve large expressions. Without more refined techniques,

the formula derivation can become time consuming and the generated code can be very long and inefficient. Thus, several problems must be solved before this approach can become widely accepted and practiced :

(i) the derivation of symbolic formulas must be made efficient and resourceful to handle the large expressions associated with practical problems,

(ii) methods must be employed to reduce the inefficiencies that are usually associated with automatically generated code, and

(iii) the system and its user interface must be designed for ease of use by engineers and scientists who have no extensive computer experience.

The system we have constructed is called FINGER (FINite element code GEneratoR) [10]. FINGER is a self-contained package written in franz LISP running under MACSYMA [6] at Kent State University. The computer used is a VAX-11/780 under Berkeley UNIX (4.2 bsd). The techniques developed are applicable not only to finite element analysis but in the general context of automatic symbolic mathematical derivation interfaced to numerical code generation.

**2.1. FINGER functionalities.** From input provided by the user, either interactively or in a file, FINGER will derive finite element characteristic arrays and generate FORTRAN code based on the derived formulas. The initial system handles the isoparametric element family. Element types include 2-D, 3-D, and shell elements in linear and nonlinear cases. The system allows easy extension to other finite element formulations. From a functional point of view, FINGER will

1. assist the user in the symbolic derivation of mathematical expressions used in finite elements, in particular the various characteristic arrays;

2. provide high-level commands for a variety of frequent and well-defined computations in finite element analysis, including linear and non-linear applications, especially for shell elements;

3. allow the mode of operation to range from interactive manual control to fully automatic;

4. generate, based on symbolic computations, FORTRAN code in a form specified by the user;

5. automatically arrange for generated FORTRAN code to compile, link and run with FORTRAN-based finite element analysis packages such as the NFAP package [1];

**2.2 Formula derivation.** In finite element analysis, the mathematical derivation leading to the material properties matrix is quite tedious and error prone. Although the manipulations involved are straightforward. By automating this process, many weeks of hard computation by hand can be avoided.

The computation involves vectors, matrices, partial differentiation, matrix multiplication etc. Expressions involved can be quite large. Thus, care must be taken to label intermediate expressions and to use symmetry relations in symbolic derivation and in generating code. The first applications of FINGER have been on elasto-plastic materials. It is interesting to note that, using the package, we have found an error (a term missing) in the plastic matrix generally accepted in the literature. In reference [2], for example, equation 12.100 in page 580 shows:

$$\frac{1}{\omega} = (1 - 2v)(2J_2 + 3\rho^2) + 9v\rho^2 +$$

(1)
$$\frac{H(1 + v)(1 - 2v)}{E} \left[2J_2 + 3\rho^2\right]^{\frac{1}{2}} \left(1 - \frac{1}{3}Bp\right)$$

Our material matrix module derived the following:

$$\frac{1}{\omega} = (1 - 2v)(2J_2 + 3\rho^2) + 9v\rho^2 +$$

(2)
$$\frac{H(1 + v)(1 - 2v)}{E} \left[2J_2 + 3\rho^2 + 2\left(r_{xy}^2 + r_{yz}^2 + r_{zx}^2\right)\right]^{\frac{1}{2}} \left(1 - \frac{1}{3}Bp\right)$$

After we found the discrepancy between the equations (1) and (2), painstaking hand computation was undertaken which verified equation (2). It is conceivable that there exist numerical finite element packages that use the incorrect material matrix formula (1).

In addition to material property matrices, FINGER also derives the element strain-displacement matrix and the element stiffness matrix in the isoparametric and hybrid-mixed formulations.

**3. FORTRAN code generation.** Actual generation of FORTRAN code from symbolic expressions or constructs is performed by the GENTRAN package [8] that we developed. It is a general purpose FORTRAN code generator/translator. It has the capability of generating control-flow constructs and complete subroutines and functions. Large expressions can be segmented into subexpressions of manageable size. Code formating routines enable reasonable output formating of the generated code. Routines are provided to facilitate the interleaving of code generation and other computations. Therefore, bits and pieces of code can be generated at different times and combined to form larger pieces. For example, consider the following sequence of steps.

1. A FORTRAN function header line is generated for the function XYZ.

2. Declarations of formal parameters of XYZ are generated.

3. Computation proceeds for the derivation and generation of the function body.

   (3.1) Some assignment statements are generated.

   (3.2) Another FORTRAN function ABC now needs to be generated (into a different output file).

   (3.3) The function ABC is generated.

   (3.4) More statements are generated for the function XYZ. Some such statements may call the function ABC.

4. The generation of XYZ completes.

The flexibility afforded by GENTRAN is evident from this example. To allow the user to control finner details of code generation and to specify the exact form of certain parts of the final code, GENTRAN allows a user-supplied "template" file to guide code generation. The template file contains literal parts and variable parts. The literal parts follow regular FORTRAN syntax. The variable parts contain code derivation and generation statements. When the template file is use to guide code generation, its literal parts stay and its variable parts are replaced by generated codes. Thus, after being processed, the template is transformed into the desired FORTRAN code. With properly specified templates, the generated code can be directly combined with existing FORTRAN code whether it's the NFAP package or something else. GENTRAN can also generate RATFOR or C code. GENTRAN has been ported to run under REDUCE [14]. The REDUCE-version of GENTRAN is available for distribution [3]. User's manuals for GENTRAN exist for both the REDUCE and MACSYMA versions.

At the present time, work is going on to add to GENTRAN an independent module to handle the generation of parallel/vectorized code.

4. **Techniques for generating efficient code.** Our experiences in automatic code generation indicate that code generated naively will be voluminous and inefficient. We have used several techniques to generate better FORTRAN code.

   (a) Automatic expression labeling: In the symbolic derivation of expressions certain intermedicate results should be generated with machine created labels. These results can be remembered to prevent the re-computation and re-generation of the same expressions in subsequent computations.

   (b) Using symmetry b generating functions and calls: Symmetries arise in practical problems and these symmetries are reflected in the mathematical formulation for solving the problem. Therefore techniques for taking advantage of symmetry are of great interest. For example, the expression $x + y - z$ is related to $x - y + z$ by symmetry, although the two can not be regarded as identical computations. If we have a function $F(x, y, z) = x + y - z$ then the latter expression is $F(x, z, y)$. If $F(x, y, z)$ is a large expression then we can simplify the resulting code generated by first generating the function definition for $F(x, y, z)$ then generate calls to $F$ with the appropriate arguments wherever $F$ or its symmetric equivalent occurs. We are not proposing an exhaustive search for symmetric patterns in large expressions. The symbolic derivation phase should preserve and use the symmetry in the given problem [9].

   This technique greatly reduces the volume of the generated code in the finite element applications. The generated code is also more structured for reading. The price to pay is the additional function calls at run time which is insignificant if the functions contain nontrivial computations.

```
t0=gl1(y,y)
t1=gl1(x,y)
t2=gl1(x,x)
sk(1,1)=(m1*t0+2*m3*t1+m6*t2)/detk
sk(1,2)=(m3*t0+m2*t1+m6*t1+m5*t2)/detk

function plpl(aa,bb)
implicit real *8(a-h,0-z)
dimension aa(4),bb(4)
v0=vi(12,aa); v1-vi(12,bb); v2=vi(10,bb); v3=vi(10,aa)
return (16.0/3.0*v0*v1+16.0/9.0*v2*v3
end

function qlpl(aa,bb)
implicit real*8(a-h,0-z)
dimension aa(4),bb(4)
v0=vi(9,aa); v1=vi(12,bb0; v2=vi(10,aa); v3=vi(10,bb)
return (-4*v0*v1-4.0/3.0*v1*v2-4.0/3.0*v0*v3-4.0/9.0*v2*v3)
end

function qlql(aa,bb)
implicit real*8(a-h,o-z)
dimension aa(4),bb(4)
v0=vi(9,aa); v1=vi(9,bb); v2=vi(10,aa); v3=vi(10,bb)
return (16.0/3.0*v0*v1+16.0/9.0*v2*v3)
end

function gl1(aa,bb)
dimension aa(4),bb(4)
return (plpl(aa,bb)+qlpl(aa,bb)+qlql(aa,bb)+qlpl(bb,aa))
end
```

FIGURE 1. Functions and calls in generated RATFOR code

Figure 1 shows generated code (in RATFOR) where functions gl1, plpl, qlql and qlpl are automatically generated with appropriate declarations. Then calls to these functions are generated to compute t0, t1 and t2 (generated labels). The function names are program generated. These functions are generated by interleaving calls to code generation routines with the formula derivation steps, resulting in great flexibility and control of the code generated.

(c) Optimizing the final expressions before code generation: Using the Horner's rule for polynomials and a search for common subexpressions for limited size expressions can help improve the efficiency of the generated code as well.

tion. Multiple windows are provided to allow concurrent control of multiple activities. A mouse is used as a pointing device to select windows and expressions, to pop up menus and to issue commands. High resolution graphics is used for mathematical symbols, fonts and interactive plotting of points, curves and surfaces. An *emacs* style editor is active whenever and wherever user input is typed. Mouse-assisted "cut and paste" allows the user to rearrange text and graphics between windows. Mathematical expressions are displayed in a textbook-like two dimensional format. Using the mouse, subexpressions of mathematical formulas can be selected interactively. User specified operations can be applied to selected subexpressions.

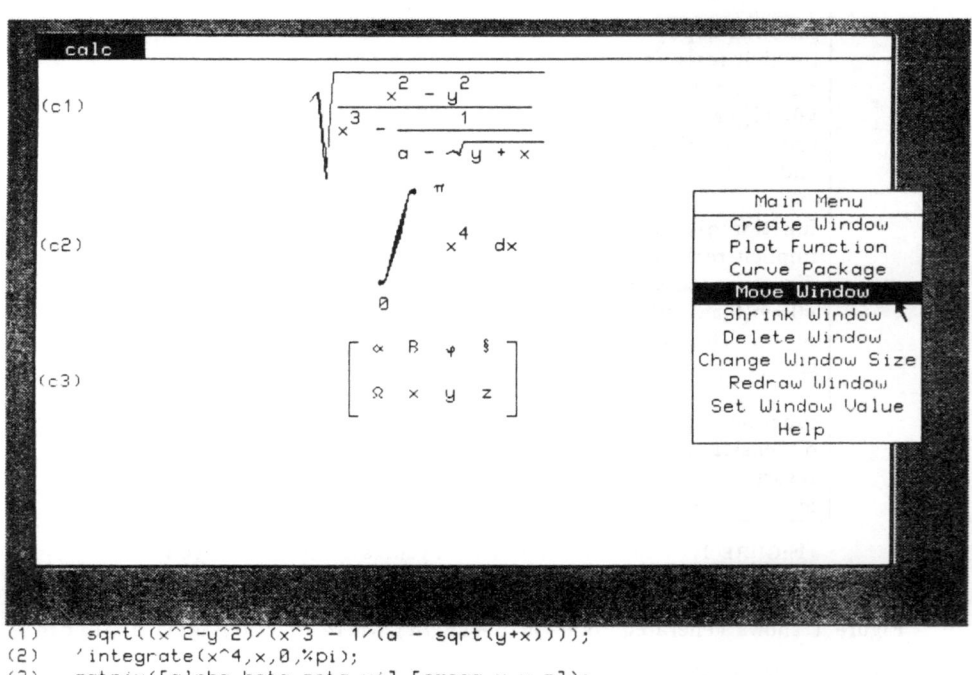

```
(1)    sqrt((x^2-y^2)/(x^3 - 1/(a - sqrt(y+x)))) ;
(2)    'integrate(x^4,x,0,%pi) ;
(3)    matrix([alpha,beta,zeta,xi],[omega,x,y,z]) ;
(4)
```

FIGURE 2.  Multiple-window user interface

**(d)** Generate subroutines: Instead of generating repeated assignments or array store operations in straight-line code, subroutines with flow control structures can be generated to reduce the code size.

5. **Graphics display for scientific computation.** Graphics display will play an important role in an integrated scientific computing environment. In such an environment graphics display should be an integral part of the user interface. A graphics package [11] for this purpose has been implemented to run under MACSYMA. This package features a highly interactive environment, a multiple window format and extensive help facilities. The capabilities include full color graphics, efficient hidden line removal, solid shading and cubic spline and least square curve fitting.

The package can display curves and surfaces given in either implicit or parametric form. The equations can be results of prior symbolic derivations. For plots involving many points, Fortran code is automatically generated to compute the function values more efficiently. The user has control over color, viewpoint, rotation, hidden line treatment etc. of plots. The control is provided alternatively through interactive menus or commands typed on the key-board. Plots can be superimposed using different colors.

The curve fitting capability allows the user to enter data points which are plotted as discreet points on the graphics display. A least square interpolation functions can then be calculated and the curve defined overlays the points. The equation for the fitted curve can be returned for further manipulation.

6. **Graphical user interfaces for scientific computing.** The user interface for a scientific computing system which combines numeric, symbolic and graphical capabilities should also be of advanced design which not only provides functionalities to control computations but is easy to learn and use. Recent studies in this direction resulted in the MathScribe [7] and the GI/S [12] user interface systems for REDUCE and MACSYMA respectively. These represent the initial steps in an investigation into suitable user interface designs for complicated scientific computing systems.

The trend is to take full advantage of the capabilities of a modern worksta-

## 6.1. GI/S windows

In the GI/S user interface system, two standard windows (figure 2) are displayed on the screen when the system begins. These are the input and display windows. The input window provides a command-line editor and a history mechanism to recall past commands. Results of computations are displayed in two dimensional form in the display window. Other windows may be opened by the user as needed. There are several different types of windows:

1. Display window

2. Scratch window

3. Graphics window, and

4. Help window.

Windows are named. Each can be relocated and re-sized interactively by the user. A corner of each output window contains status information on how the computation controlled by the window is progressing. The mouse buttons are used for selection and for appropriate pop-up menus.

## 6.2 Mouse apply

One way to exploit the capability of the mouse in a scientific system is to use it to enhance mathematical operations. One such operation is singling out a part of a large expression and apply a user specified function to it with the result of the function replacing the original part *in place*. Let us call this operation "mouse apply". Figure 3 and figure 4 show the "before and after" of *mouse applying* the function **factor**.

Studies of user interface design of complicated scientific systems have just begun. Standards, protocols and conventions are still largely lacking. However, one can be sure that advances will be made and users will benefit much from the next generation interface systems.

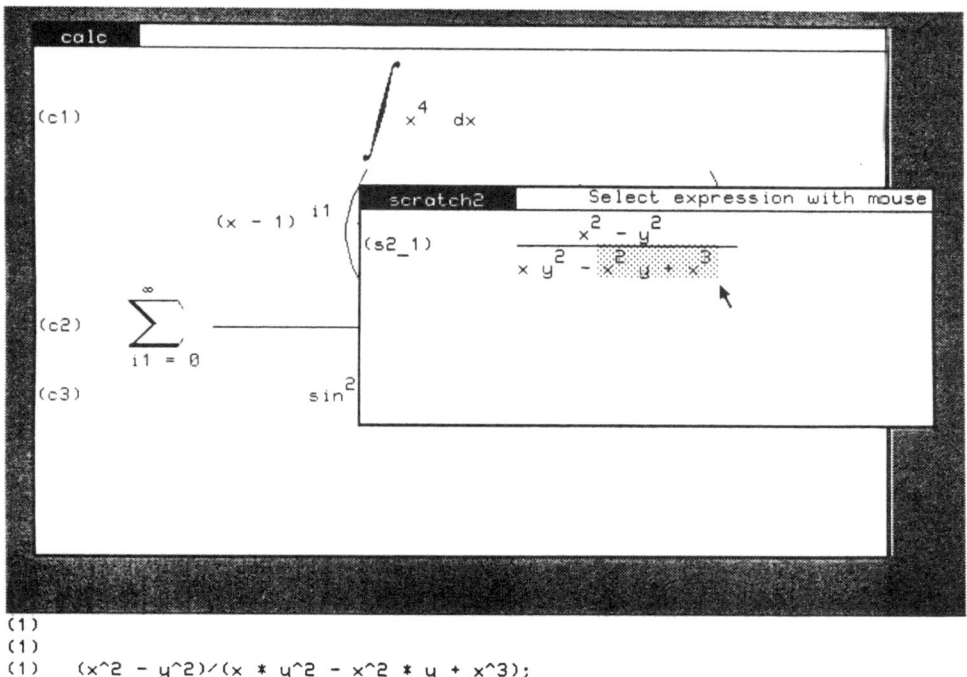

FIGURE 3. Before *mouse-apply* factor

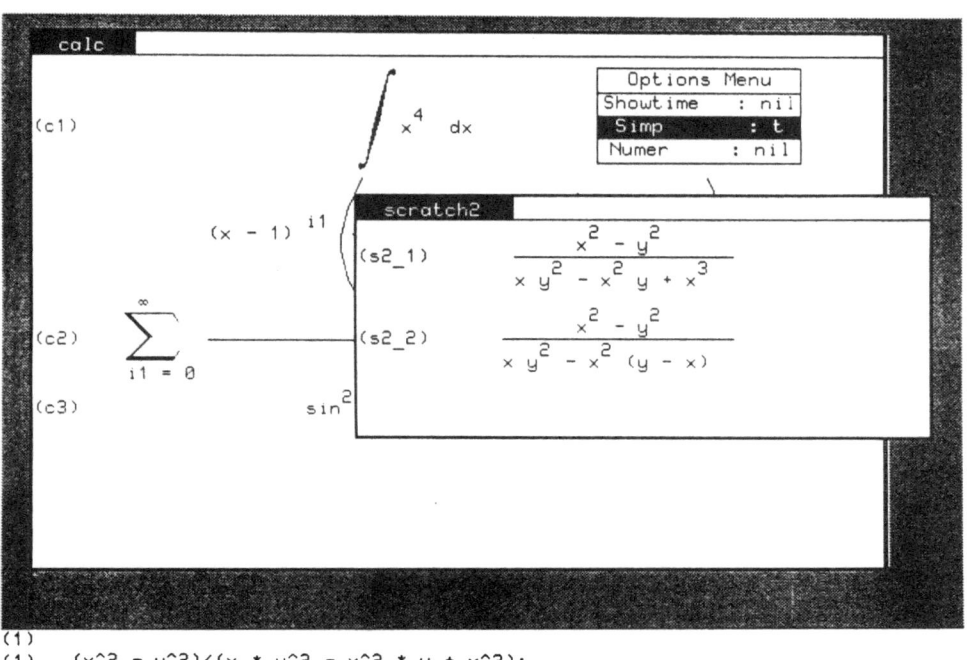

FIGURE 4. After *mouse-apply* factor

**7. Conclusions** Modern workstations offer a practical way to integrate numeric, symbolic and graphics computing systems into one comprehensive scientific computing environment. Operations such as symbolic formula derivation, automatic numerical program generation, graphics display of data points and mathematical equations, and advanced user interfaces can work together and offer many desirable features and capabilities that are otherwise unavailable. Evolution of such integrated environment will one day provide a powerful tool for scientists and engineers for substantially increased productivity.

## REFERENCES

[1 ] CHANG, T. Y., NFAP - A Nonlinear Finite Element Analysis Program, Vol. 2 - User's Manual. Technical Report, College of Engineering, University of Akron, Akron Ohio, USA 1980.

[2 ] CHEN, W. F., *Limit Analysis and Soil Plasticity*, Elsevier Scientific Publishing Co., New York, 1975.

[3 ] GATES, B. L., "GENTRAN: An Automatic Code Generation Facility for REDUCE",

[4 ] KORNCOFF, A. R., FENVES, S. J., "Symbolic generation of finite element stiffness matrices", Comput. Structures, 10, 1979, pp. 119-124.

[5 ] NOOR, A. K., ANDERSEN C. M., "Computerized Symbolic Manipulation in Nonlinear Finite Element Analysis", Comput. Structures 13, 1981, pp. 379-403.

[6 ] PAVELLE, R. AND WANG, P. S., "MACSYMA from F to G", Journal of Symbolic Computation, vol. 1, 1985, pp. 69-100, Academic Press.

[7 ] SMITH, C. J., SOIFFER, N., "MathScribe: A User Interface for Computer Algebra Systems," Proceedings, the 1986 Symposium on Symbolic and Algebraic Computation, 1986, pp. 7-12.

[8 ] WANG, P. S. AND GATES B., "A LISP-based RATFOR Code Generator", Proceedings, the Third MACSYMA Users Conference, August, 1984, pp. 319-329.

[9 ] WANG, PAUL S., "Taking Advantage of Symmetry in the Automatic Generation of Numerical Programs for Finite Element Analysis", Proceedings, ACM EUROCAL'85 Conference, April 1-3 1985, Lecture Notes in Computer Science No. 204 (1985), Springer-Verlag, pp. 572-582.

[10 ] WANG, P. S., "FINGER: A Symbolic System for Automatic Generation of Numerical Programs in Finite Element Analysis", Journal of Symbolic Computation, vol. 2, 1986, pp. 305-316, Academic Press.

[11 ] YOUNG D. A. AND WANG, P. S., QAn Improved Plotting Package for VAXIMA", abstract, presented at ACM EUROCAL'85 Conference, April 1-3 1985, Linz Austria, Lecture Notes in Computer Science No. 204 (1985), Springer-Verlag, pp. 431-432.

[12 ] YOUNG D. A. AND WANG, P. S., "GI/S: A Graphical User Interface For Symbolic Computation Systems", Journal of Symbolic Computation, Academic Press. (to appear in 1987).

[13 ] *COSMIC NASTRAN USER's Manual*, Computer Services, University of Georgia, USA.

[14 ] *REDUCE User's Manual*, Version 3.0, Edited by Anthony C. Hearn, The Rand Corporation, Santa Monica, California. April 1983.